THINGS
to do with
PLANTS

THINGS
to do with
PLANTS

50 ways to connect with the botanical world

Emma Crawforth

.

Kew Publishing
Royal Botanic Gardens, Kew

First published in 2022 by
Royal Botanic Gardens, Kew,
Richmond, Surrey, TW9 3AB, UK
www.kew.org

ISBN 978 1 84246 779 4

Distributed on behalf of the Royal Botanic Gardens, Kew in North America by the University of Chicago Press, 1427 East 60th St, Chicago, IL 60637, USA.

British Library Cataloguing in Publication Data
A catalogue record for this book is available from the British Library

DESIGN: Kevin Knight
PROJECT MANAGER: Lydia White
PRODUCTION MANAGER: Georgie Hills
COPY-EDITING: Matthew Seal
PROOFREADING: James Kingsland

For information or to purchase all Kew titles please visit shop.kew.org/kewbooksonline or email publishing@kew.org

Kew's mission is to understand and protect plants and fungi, for the wellbeing of people and the future of all life on Earth.

Kew receives approximately one third of its funding from Government through the Department for Environment, Food and Rural Affairs (Defra). All other funding needed to support Kew's vital work comes from members, foundations, donors and commercial activities, including book sales.

Printed and bound in Germany by Firmengruppe Appl

Picture credits: Tony Hall: 33; RBG Kew: 38, 55, 57, 62, 108, 109, 113, 114; Shutterstock: 72, 75.

Contents

Introduction

We all have a relationship with plants, members of a vast kingdom of organisms, consisting of 390,900 species at the last count. That relationship varies from person to person, according to knowledge and experience, but everyone shares a reliance on plants to make food using the light of the sun and now, more essential than ever, to save us from the terrible dangers posed by climate change.

Given how important plants are to us, it's always surprising to see how little attention they receive. There's a lack of curiosity that's been termed 'plant blindness'. Plants are rarely in the news, recognised in culture or even studied much in their own right, often being treated as just part of 'biology'. As you've picked up this book, I'm assuming that you, like me, are curious about plants.

I've been working with and studying plants full time for twenty years now, firstly as a horticultural student, then as a gardener and finally as a gardening journalist. I'm interested in all plants, never specialising in one group, and I'm prone to go off on a tangent once I've discovered something new, I hope you enjoy journeying along those tangents with me.

Writing this book gave me the opportunity to visit and revisit things to do with plants that I was particularly curious about. The title is deliberately ambiguous, representing both information about plants and things that we humans actively do with them. The list of things to do is far from comprehensive and focuses on activities that intrigue me. I've thoroughly enjoyed researching some of the ways in which plants are transformed into objects we use in our everyday lives.

Plants have a huge capacity to enhance our everyday lives. Just consider how different this pavement would be without them

The book is divided into several parts, from the universal to the personal. The universal focus covers the global influence of plants, their role in making this planet a habitable place, where food, clean water, medicine and air to breathe are so commonplace for most of us that we barely think about them.

The community activities are focused on the things we do together once our basic conditions for survival are met, organising society, building houses, working and celebrating as we live together.

Finally, I've covered more personal things to do with plants. These are activities we can all try alone, although many of them are more commonly done collectively. It's a canter through the basics of using plants to enhance our everyday lives, and I hope you'll be intrigued enough to have a go at doing some of these things yourself.

There is often a synthetic alternative to be used instead of a plant, but that alternative gives none of the benefits of carbon capture, eco-friendly disposal and sheer attractiveness conferred by plants. I also think that working on some of these activities with plants is very satisfying and provides a good distraction from many more destructive activities. For example, anyone who is obsessed with their garden finds it very hard to fly off for long holidays abroad!

Of course, plants are just one part of the world's extraordinary ecosystem, a complex web that is damaged as soon as it's unbalanced. That's why I'm focusing on using plants in ecologically sound ways. We can exploit them by growing them in monocultures supported by unnatural chemicals to improve profits through speed of production, or we can treat them with respect, retaining the integrity of the ecosystems to which they belong and enjoying the benefits that diversity provides. The point is not just neutral, as exploitation of plants can harm our world and we need them to help us heal it more than ever.

When pondering how to tie together all of the things to do with plants I'd written about, I realised that their relationship could easily be represented by the form of a tree, with the universal things making up the trunk, community activities the branches and finally personal ones, the twigs. How apt that once again the kingdom of plants had solved my problem!

Gardens designed to support wildlife like this one improve our environment while boosting our sense of wellbeing

Save the world!

Slow climate change

Human-made climate change is caused by the release of certain types of gas into the atmosphere, the dominant one being carbon dioxide. Climate change has led to wildfires, droughts, intense heatwaves, flooding, sea-level rise from loss of sea ice, shrunken glaciers and disastrous shifting of plant and animal natural ranges. The progress of climate change, along with these catastrophic events, is non-linear rather than steady.

Growing plants reduce atmospheric carbon by photosynthesis. They take carbon dioxide (CO_2) from the air and convert the carbon into the complex molecules from which they are made – proteins, fats

and carbohydrates. They also (like us) respire, taking in oxygen and releasing carbon dioxide. Overall, however, growing plants take in far more carbon than they release. A few other types of organism – phytoplankton, lichens, cyanobacteria and one or two animals – can do it too, but it's plants that are doing the most important work.

While living plants capture carbon, dying plants gradually release it. As animals and other organisms break down the carbon-containing compounds from which plants are made, those organisms respire and release carbon back into the atmosphere. So, maintaining environments where carbon is stored safely in plant matter (living or dead) is vital.

Waterlogged peatlands are thought to contain up to a third of the world's soil carbon, on just 3 per cent of the Earth's surface. This

Forests not only sequester carbon but they also replenish clouds by releasing water from their foliage

ACTION PLAN TO WORK WITH PLANTS AGAINST CLIMATE CHANGE

This is a list of easy actions we can take to decrease our own carbon footprints using plants and gardens:

❉ Grow a garden full of plants rather than paving it over for easy maintenance and car parking

❉ Grow and harvest our own vegetables and cut flowers rather than buying them from shops with a large carbon output

❉ Avoid using peat-based products

❉ Compost any plants cut from our gardens and use the compost to enrich our soil to support the plants growing in it (see Green up a garden, page 80)

❉ Protect rainforests by rejecting the products of their destruction, such as palm oil, an ingredient of processed foods and toiletries

❉ Support responsible re-forestry projects run by charities such as Cool Earth, Woodland Trust, World Land Trust and World Wide Fund for Nature (WWF)

carbon is held within dead plant matter protected under water, but if that peat is dug up and used for cooking, heating or gardening, its breakdown releases carbon into the atmosphere quickly.

Trees, of course, are vital for our global effort to control climate change. According to carbon expert Mike Berners-Lee, each deforested hectare is equivalent to driving a car 50–100 times around the world (a hectare is roughly 1.5 football pitches). Forests currently take up about 25 per cent of the world's land area, but an area of forest similar to the size of Switzerland is lost annually.

Each person in the UK has an average carbon footprint of around 13 tonnes per year, and planting 100 square metres (120 square yards) of broadleaf forest in the UK could remove just 5 tonnes of atmospheric carbon per year. That's a lot of tree planting to counter our individual use of carbon, which is why the solutions to this problem need to be holistic (see the action plan opposite).

When it comes to carbon capture, one of the hardest-working groups is the seagrasses (flowering plants that grow in oceans) – these colonise shallow areas and are thought to absorb 10 per cent of the ocean's carbon each year through photosynthesis and trapping floating particles, so they're vitally important in the fight against climate change. There are many different species, but the most effective could trap carbon 35 times faster than rainforest!

The world's largest individual plant is an Australian seagrass (*Posidonia australis*) which covers an area bigger than Washington DC. Seagrasses also grow around the British Isles and can be seen on Thames estuary mud flats, but unfortunately many populations in our estuaries have been reduced during industrialisation.

It's staggering to think that while plants are the innocent victims of our irresponsible behaviour as their habitats are ruined, they can also be our saviours.

Restore natural landscapes

It seems obvious that to restore a landscape you'll need to grow plants, because all natural landscapes on Earth, apart from the most extreme and inhospitable, include plants. But the plants in landscapes are much more than idle constituents. They control soil erosion, serve as habitat for wildlife and purify water. These are the building blocks for ecosystems. Once an ecosystem is restored to good health it will provide economic and social benefits, reduce extinctions and clean the air as well as slowing climate change by storing carbon.

Restoration projects aim to improve degraded ecosystems so they can function once more and support a diversity of living creatures. They can restore or repair a degraded natural ecosystem or improve land in human use, for example by planting hedgerows to boost soil quality on farmland. Often, it's depleted forests that are targeted.

It's thought that over 50 per cent of the Earth's original forest has gone and there are more than two billion hectares of degraded land where intervention could work. This is particularly important because forests are a crucial weapon in the fight against climate change.

For example, in north-west Pakistan a project to add 350,000 hectares (865,000 acres) of trees and improve 160,000 hectares (395,000 acres) of degraded forest is expected to result in the sequestration of 1.7 million tonnes of additional carbon each year. A network of private tree nurseries was created across the region, giving jobs to local

people, including women and unemployed young people, and the project was deemed a success by an independent World Wide Fund for Nature (WWF) audit.

The UN Decade on Ecosystem Restoration began in 2021 with projects aimed at savannahs and shrublands, rainforests, peatlands, urban spaces and, in the UK, seagrass and peatland restoration plus the regreening of Manchester.

And even if these projects still seem remote from our everyday lives, we can all play a part by:
• Holding our governments to account for their promises – the UK has pledged to restore 381,000 hectares (941,000 acres) of degraded land
• Supporting charities, such as RBG Kew and WWF, with carrying out restoration projects
• Controlling our consumption of products that lead to degradation, like exotic timbers, fast-fashion clothing and peat-based garden composts.

Stabilise soil

Human activities have increased the global rate of soil erosion by 10–50 times. In the UK alone it's estimated that 2.2 million tonnes of topsoil are eroded annually. Why does that matter?

Soil erosion decreases agricultural productivity and causes ecological collapse when the nutrient-rich upper soil levels are lost. As the soil is displaced eutrophication occurs (that is, algal growth as a result of an accumulation of nutrients and minerals) and waterways are blocked. Sometimes poisonous chemicals are transported along with the soil.

Climate change, intensive agriculture, deforestation and urbanisation cause soil erosion, usually by exposing soil to water and wind, which transport it away from the areas where it's needed. From an island, it can simply be washed out to sea.

Just as removing trees in deforestation causes soil erosion, so planting trees is a remediation method. Shelterbelts are windbreaks made of trees to shield land. They don't just filter wind, they can also improve the land's microclimate, reducing dehydration, for example.

On tropical coasts, mangroves (salt-tolerant coastal thickets) with their complex root structures reduce wave and flood damage from storms while they bind soil together. Mangroves are considered a powerful weapon against damage from tsunamis.

While mangrove species protect soil from sea water, grasses reduce erosion from coastal winds. Sand dunes founded on grasses stop beach erosion, acting as a barrier against flooding of inland areas

This pine tree is growing in a spot where weather quickly erodes bare earth, but its complex root system is protecting the soil and providing a habitat for rare and endangered animals and plants. They are made up of sand held together by specially adapted dune vegetation, which is tolerant of desiccation, salt and changing temperatures, to name a few challenges.

The delicate habitats in coastal dunes are easily disrupted, so restoration projects involve native plants. In the UK, marram grass (*Calamagrostis arenaria*) is a key species as its fast-growing rhizomes form a good anchor in sand. It tolerates salty air and water and even movement as the dunes shift around.

Sophisticated landscape restoration projects may appear difficult for ordinary plant lovers to relate to, but we can all help to fight soil erosion. There are coastal restoration projects run by wildlife organisations like The Wildlife Trusts in the UK that volunteers can take part in, and when we visit the beach, sticking to walkways over dunes is essential.

Shrimp farms are considered a major threat to mangrove areas, so we should aim to buy sustainable fish. Similarly, intensive agriculture results in soil erosion from the large open fields, so choosing groceries from ethical sources helps.

Mangrove species protect seashores from erosion by waves, and sand-dune plants, like marram grass, protect them from wind erosion

This ancient and beautiful wisteria in Kew Gardens has protected visitors sitting beneath it from the sun's heat for many years

Create shade

Gardeners in the UK often complain about shade cast in their gardens, but deliberate shade creation is becoming more and more important as climate change bites. This need is particularly acute in urban heat islands, which occur in cities where hard surfaces magnify the effect of warmth from the Sun and human activity creates yet more heat. Statistics show fatalities in vulnerable groups, such as the elderly, are higher in these areas during heatwaves.

We turn to air conditioning, making the problem worse by increasing carbon emissions, but we can improve the situation by making more summer shade in our plots. All gardens, wherever they are, need shady areas, to provide sanctuary from summer heat, protection from harmful radiation, a habitat for wildlife and a cool spot that will suit a pond or chicken coop.

Planting trees or shrubs and growing climbers on a structure are easy ways to bring shade into your garden. Before you start digging it's a good idea to make plans, though. Create a sun map of your garden by taking photos from an upstairs window at different times of the day and through different seasons, then superimpose your photos digitally to get a complete picture. Alternatively, you can download apps that will do this for you.

Consider the impact that a tall structure would have on the places where shade would fall. In a temperate climate, you probably only want shade in the summer, so think about using deciduous or annual plants.

Research by Reading University and the Royal Horticultural Society highlights the benefits of growing climbers like ivy directly on houses. They found that not only did this cool summer temperatures, it also (contrary to general beliefs) reduced humidity.

If you're going to plant a garden tree, choose one that grows quickly, so you can enjoy the effects yourself. If you like a formal look, pruning in a parasol shape is possible with some species. Good garden trees include birches, alders, rowans, hazels and beeches. These also make robust hedges (which are great for garden wildlife) if planted close together and trimmed regularly.

All these species are deciduous, giving seasonal change as the leaves come and go. When choosing trees, always remember the 'right plant, right place' mantra of the well-known gardener Beth Chatto,

ACTION PLAN FOR CREATING SHADE

❋ Make a list of the reasons why you want a shady spot
❋ Estimate the area where you want to cast shade
❋ Work out the best size and shape of structure or plant to use
❋ If using a structure, plan where its foundations would be dug
❋ Research the trees, shrubs, climbers or annuals that will grow to the desired height and spread
❋ Double check that the plants will thrive in the spot where they need to grow
❋ Choose shade-loving plants to grow underneath.

as picking one that won't thrive is harmful to your peace of mind, pocket, garden and sustainability.

Alternatively, robust pergolas, arches and bowers will support climbers such as roses, wisteria and clematis. Consider planting two different ones on the same structure, to get a longer season of flowering as each plant blooms in turn. With climbers you'll need to understand their pruning requirements – this can make the difference between a beautiful garden feature and a tangled mess!

Don't restrict yourself to perennial plants, though. Climbing annuals cast shade when you need it in summer and die off in time to bring winter light into your plot. A 'green screen' designed by Japanese architects Hideo Kumaki is a great example of this approach. The temporary net is angled towards a low roof at 45 degrees for annual legumes and morning glories to climb, casting shade from July to October. A 10°C (50°F) temperature difference has been measured between the sunny outside and shady inside of the screen.

Finally, plan the groundcover planting you'll grow beneath your shade-givers. This is easier under deciduous plants as many shade-lovers are fine with just winter sun. Cyclamens, snowdrops, hellebores, primulas and many ferns will thrive in a high-winter, low-summer light regime.

Grow protein

Protein is an essential part of the human diet, as although our bodies can synthesise some of the 20 amino acids (protein building blocks) we need, nine of them come only from our food. From our hair to our toenails, via brain, heart and skin, we are partly made up of protein – it's an estimated 17 per cent of our body weight. It helps maintain our immune system, blood sugar regulation, fat metabolism and energy function. Proteinaceous foods also supply important vitamins and minerals.

Fortunately, it's surprisingly easy as gardeners to grow our own protein, with some of the most common vegetables, such as garden peas, kale and sweetcorn, containing high amounts.

In a temperate climate, the ideal protein garden contains trees, climbers and groundcover, with crops for harvesting all year round. But summer is the high season and the time when you need to start picking for storage, so you have plenty of protein-rich ingredients for winter cooking.

Vegetable gardens benefit from crop rotation, where annual crops are moved around the garden, to deter pests and diseases as well as conferring benefits on the soil – legumes leave nitrogen behind that brassicas can take up, so a protein garden should be a dynamic place that changes from year to year.

However, I recommend starting with the most static element – a hazelnut tree. Not only are nuts a great protein source, but a hazel will give you year-round beauty, starting with winter catkins, and

benefits including shade, wildlife cover and a supply of supports for your climbers. Coppiced hazel twigs make the best pea and bean sticks. Cut them from the base in winter, to fashion into arches and tripods.

The other edible perennial I recommend is asparagus. The spears are surprisingly high in protein. It takes a little while to establish but is easy to care for after that and the delicious crop appears during the 'hungry gap' in April–May – the quietest time in the vegetable garden.

Amaranth seed is high in lysine, one of the building blocks of protein, which is essential in the human diet

Pumpkin seeds contain several different amino acids, making them a useful ingredient for a protein-rich meal plan

Peas, beans and pumpkins can be grown on your hazel supports. Not all of them need support, but by growing them up, rather than along, you save soil space for more crops. The pulses – shelling peas and broad beans – are hardier climbers. In fact, you can sow them in autumn to establish over winter and grow away in spring. Sow the pumpkins in late spring, to harvest for their protein-rich seeds in October.

For succulent winter harvests, grow the brassicas – Brussels sprouts and kale. These need sowing in spring and will quietly grow tall while you're tending everything else. They may need a little support to avoid leaning.

Sow amaranth and quinoa to produce masses of seeds you'll be able to store and eat over winter. Quinoa contains all of the 20 amino acids we need, so although there's a little effort involved, the rewards are high.

To harvest amaranth, rub the flowers between your hands and check that plenty of seeds are being released. This should be done before the flowers have dried out. Strip off the flowers and sieve over a bucket then spread out the seed to dry. Next, winnow by pouring it between bowls in a light breeze. Dry it again and store in an airtight container.

The process for quinoa is similar but cut the stems when the seeds are changing colour and hang to start drying before rubbing between your hands. Quinoa seed also needs drying and winnowing. Other crops to store dry are podded beans, peas, sweetcorn, pumpkin seeds and hazelnuts.

For summer bounty sow sweetcorn in mid-spring, spinach in succession every few weeks, from mid spring, runner and French

MAKING A PROTEIN GARDEN
– WHAT TO DO WHEN

❄ JANUARY–FEBRUARY: plant hazel, harvest Brussels sprouts and kale, cut hazel twigs

❄ MARCH: sow Brussels sprouts and kale, plant asparagus

❄ APRIL: sow quinoa, amaranth, sunflowers and sweetcorn under cover, harvest asparagus, start sowing spinach and continue every six weeks until September

❄ MAY: sow pumpkins, French and runner beans under cover, harvest broadbeans

❄ JUNE: plant out amaranth, quinoa, sweetcorn, French and runner beans

❄ JULY–AUGUST: harvest peas, beans, sweetcorn and spinach

❄ SEPTEMBER: harvest quinoa

❄ OCTOBER: harvest hazelnuts, pumpkins, sunflower seed, amaranth (best after a light frost)

❄ NOVEMBER–DECEMBER: cook and eat dried beans, hazelnuts, quinoa and amaranth, plus pumpkins, kale and Brussels sprouts.

beans. Sunflowers will make edible seeds in a hot summer when planted in a sunny spot. They add colour and support wildlife too.

Finally, many of these vegetables are easy to grow as sprouting seeds, which are a great source of protein. Take care to use seed that has not been treated with chemicals and research your seeds before you start, as some beans, for example, contain toxins when raw.

Filter water

For many of the chapters in this book I'd recommend you have a go at the projects without hesitation. In this case, getting the method wrong could have serious health implications, so if you're going to try any of these methods, you'll need to do plenty of your own research. However, it's fascinating to discover how many different types of plant-based systems there are for purifying water. They rely on physics, chemistry and of course biology.

The most well-known method in the UK must be the reed-bed filtration system. Wastewater is passed through a bed of reeds, such as *Phragmites australis*. The water moves from one end to the other of a planted basin, while oxygen given off by the reed roots is used by bacteria that process harmful chemicals and microbes.

There are many different versions of this system, taking up different amounts of land space and producing various grades of cleaner water. Some can be used to irrigate crops, for example. Off-grid homes welcome these systems as no energy is needed during their day-to-day running. Some use this method simply to process grey water (such as washing waste) but others even process sewage water like this.

One fascinating filtration method uses a plant's own water pipes (xylem) to clean up water. Researchers from the Massachusetts Institute of Technology tightened a plastic tube over the end of a white pine (*Pinus strobus*) branch. They discovered that water flowing through the freshly cut sapwood was cleansed of pathogenic bacteria, due to the tiny pore size within the wood.

Pollution from heavy metals can also be countered by plants. The popular herb coriander/cilantro (*Coriandrum sativum*) has proved effective at removing toxic metals like lead and arsenic when suspended, teabag-like, in contaminated water. Apple and tomato peel as well as certain mosses are said to have a similar action. Coconut shell fibres with rice husks (or banana peel) are employed to physically filter solids from water. The popular garden plant mahonia contains berberine, an antimicrobial alkaloid, in its inner bark, making it the friend of North American hikers stranded in the wild.

Finally, a plug for limes (*Citrus* x *aurantiifolia*), which speed up the inactivation of *E. coli* bacteria when water is cleaned using the SODIS method, which involves leaving water in sunlight to prevent multiplication of the bacteria. It is the psoralens (chemical compounds that change DNA) in the limes that are effective.

A tea made from mahonia bark, a source of antimicrobial berberine, is thought to aid adventurers who have drunk impure water

Make medicine

Many of the most commonly used drugs in modern medicine are either directly extracted from plants or synthesised copies of plant substances.

For example, the yam (*Dioscorea mexicana*) is a source of steroids, which were modified to produce oestrogen for birth-control pills. A group of pretty, bulbous plants – daffodils, snowdrops and amaryllis – are a source of galantamine, used to treat Alzheimer's disease. The name is related to the scientific name for snowdrops, *Galanthus*.

Yew trees, *Taxus brevifolia* and *T. baccata*, were the source of the drug Taxol for treating breast and ovarian cancer (discovered by research workers in North Carolina from an extract of yew bark), and Madagascar periwinkle (*Catharanthus roseus*) improved the chances of survival of a child with leukaemia from 20 to 75 per cent thanks to its extracts vincristine and vinblastine.

The plants I've mentioned were commonly used as remedies before modern medicine found them. Often, as they contain strong poisons, it was on a kill-or-cure basis! Certain plant groups use 'secondary compounds' to help them deter predators, inhibit the growth of competing plant species or attract pollinators, and these plants tend to be the ones with medicinal value too.

But while the traditional use of these plants provided clues for doctors, we have only discovered the tip of the iceberg when it comes to potential plant-based cures. Out of 390,900 plants known to science, it's estimated that only 1,150 have been investigated

Two chemotherapy drugs were developed from chemicals called taxanes originally found in *Taxus* (yew) trees for their medicinal properties, and pharmaceutical companies are now combing the South American rainforests for plants containing medicinal compounds.

You don't have to search the rainforests for healing plants, though. In an ordinary British garden, for example, you could be growing many of the plants thought to help with one of the most common modern-day health complaints – insomnia.

It's a complex problem, with many causes and several treatments, but the effectiveness of some well-known garden plants has been proved in trials. Many of these plants contain melatonin, a hormone that occurs naturally in the body and helps control sleep patterns. Melatonin tablets or drinks are prescribed to help people fall asleep faster, stay asleep during the night and cope with jetlag symptoms.

YEW

CHERRY

Kiwi fruits and sour cherries, both sources of melatonin, have proved good for sleep in tests. Lettuce, traditionally associated with sleep, has also fared well and contains lactucin, which has sedative qualities. (Beatrix Potter in *The Tale of the Flopsy Bunnies* mentions soporific lettuce!)

While fruit and salad cures would all be eaten, it seems lavender only has to be sniffed to assist sleep. Recent randomised studies showed good results from lavender odour on insomnia in midlife women, heart-disease patients and healthy students alike.

Some of the oldest known gardens in the world are 'physic' gardens, places where herbs are grown for their medicinal benefits. They have a rich history, but it looks as though we need them just as much now as ever before.

LETTUCE

LAVENDER

Build a community

Celebrate lives and occasions

Planting a tree to celebrate the birth of a baby is common across many cultures and is a wonderful way to mark a new child's entrance into the world. Some mothers even bury the placenta under the sapling, to nourish it with the nitrogen contained in the placenta blood.

If that child grows to adulthood and marries, then plants will almost certainly be part of the celebrations. In the UK a bride holds a bouquet, often of white flowers, while in Indonesia it is white melati blooms (*Jasminum sambac*) that adorn the groom's ceremonial

dagger and the bride's hair. Henna decorations are painted on a bride's skin in Asia, Africa and the Middle East. They are made from the powdered leaves of the tree *Lawsonia inermis* (see page 40). The couple may also be showered with grains of rice, thought to symbolise prosperity and fertility.

Not all cultures mark the end of life with festivities, but on the Mexican Dia de los Muertas (Day of the Dead), the lives of deceased loved ones are celebrated through the creation of vibrant altars with vases and garlands of the flower of the dead, marigolds (*Tagetes erecta*). This orange bloom is a popular bedding plant in the UK but it's native to Mexico and is used to encourage visits from the spirits of the dead with its bright colour and pungent scent.

Bright hues also mark the Hindu festival of colours, Holi, which heralds the arrival of spring with singing, dancing and a liberal

Groves of cherry trees provide a perfect spot to celebrate the end of winter during the festival of Hanami

How to make patterns on the skin with henna

1 Use a fine powder of *Lawsonia inermis* leaves

2 Mix it with a liquid such as water, lemon juice or tea

3 Try adding sugar to make the paste stick better and lavender or tea tree oil to improve staining

4 Rest the paste for up to two days so the staining chemical, lawsone, is released

5 Apply the paste with a stick or icing bag

6 Leave on the skin for 4–12 hours

7 Brush the paste off or loosen with vegetable oil

8 Watch as the dye develops from orange to red–brown over the next three days

9 Allow the mixture to gradually wear off as the skin exfoliates. It will last longer on the thick skin of feet and hands.

Henna dyes are pale when first applied but the pigment darkens through exposure to air after a couple of days

sprinkling of coloured powders. Natural, plant-based sources of colour include turmeric, beetroot, sandalwood, henna, spinach, coriander, saffron and jumun (myrtle).

Also celebrating spring, but with pastel colours, is the Japanese Hanami festival, when the cherry blossom, or Sakura, is enjoyed during family gatherings, picnics and parties. Groves of Japanese cherries such as *Prunus serrulata* form the setting, with the best timing for gatherings predicted in blossom forecasts.

At the other end of the year, red holly berries brighten the shortest and darkest days in Western Europe as they're strung up in garlands and door wreaths. Meanwhile in Eastern Europe an oak (*Quercus cerris*) sapling may be cut at Christmas, forming the Badnjak log. Its burning is accompanied by prayers that the coming year will bring happiness, riches and luck.

Tagetes erecta **has many jolly cultivars and is called African marigold, although it's native to Central America**

Investigate an archaeological site

I've met several gardeners with a curiosity about archaeology. A link, through digging and the soil, seems obvious while a broad understanding of the natural world underpins both interests. But whether you know much about plants or not, they can give helpful clues to the lives of our ancestors.

In 2018, a summer of severe drought in Ireland, an ancient monument dating back 5,000 years was discovered in a wheat field. Photographs from a drone flying over the field clearly showed a circle indicating the presence of a henge (circular earthwork) that had once stood on the land. This would have enclosed a monument similar to Stonehenge and Avebury in England, but made of wood. The circle had never been noticed before (in modern times) and disappeared when the drought ended.

The simple explanation for this Neolithic, ghost-like feature is that the foundations of the wooden posts, buried underground, changed the soil. Organic matter added to soil helps it to retain more water, so in times of drought, like the summer of 2018, the wheat did better when growing over the footprint of the henge than elsewhere in the field, making a distinct impression. This is not an unusual story as archaeologists routinely look for changes in the colour or texture of vegetation to indicate something happening underground.

To work in palynology (the study of pollen, spores and plankton) you need an extensive botanical background. Archaeologists

use palynology to discover what was growing on a site – crops, ornamentals or wild vegetation – in ancient times. Microscopic pollen grains and spores survive in soil for thousands of years (due to their dense structure) when little oxygen and physical disturbance are present. Radiocarbon data can determine their age.

The genus or even species of a plant can be identified through its pollen (spores from fungi are slightly less helpful). Because wind-pollinated species, like grasses and conifers, produce more pollen than insect-pollinated ones, they are well represented in archaeological records.

You don't always need a microscope to discover how plants were used in ancient times though, as in extremely dry places whole dried flowers have been found, like the garlands of cornflowers (*Centaurea cyanus*) and mayweed (*Anthemis pseudocotula*) discovered in the tomb of Tutankhamun.

Dried garlands of cornflowers found in a tomb helped archaeologists discover more about life in Ancient Egypt

Gain a quieter environment by planting a tall and wide hedge, like this one of common yew

Reduce noise

Sounds can be beautiful, but noise is generally annoying and, more seriously, it causes tension, anxiety and sometimes even hearing loss. Prolonged noise levels of 40–65 decibels are considered harmful to human health, so research showing that belts of vegetation can reduce noise by up to 8 decibels is welcome. According to the Forestry Commission, this equates to a reduction in noise to the human ear of approximately 50 per cent, due to the way the decibel scale works.

Vegetation reduces noise by sound attenuation or damping, that is, diminishing the volume and quality of a sound wave. All plant parts absorb and deflect sound energy. Some of the absorbed energy is transformed into heat and lost. Sound waves can be refracted, as they bend around plant structures, too. Vegetation also masks noise, by making its own pleasing rustling and creaking sounds, or attracting wildlife, such as songbirds.

To reduce noise most effectively, a vegetation barrier should not have gaps, hence hedges are very effective. Barriers should start at ground level and be tall (at least 2m), with close planting, ideally of evergreen shrubs or trees, so the effect lasts all year round.

The Royal Horticultural Society recommends cherry laurel, yew, holly, berberis and thuja for an effective noise-reduction hedge. Clipping will increase the density of the foliage, and hence improve the sound barrier over time. And for maximum effectiveness the hedge should be planted close to the source of noise rather than the area to be protected.

Mark a route

There is a rich collection of tree avenues in the world, each one
changing the landscape in an impressive way, providing shelter
and a sensory experience of cool shade. The most renowned
are planted with one tree type, giving them an identity through
uniformity, with the individual trees performing the same way
throughout the seasons and growing into matching forms. But
this way of planting will change in the future, because using a
diversity of species is key to tree health.

Route marking is not always done by planting new trees as
established ones can simply be painted to show trail paths. This is a

modern expression of a technique going back into the mists of time of breaking or notching a branch at regular intervals on a journey as a guide for the return trip.

Perhaps that custom led to deliberate planting along routes. Empress Maria Theresa of the Habsburg Empire in 1752 ordered the planting of trees on new roads (prescribing the spacing between the trees – 11.4 m (37 ft) – and the quality of the saplings) partly to increase wood production and for aesthetic reasons, but also to help travellers orient themselves through snow and fog. Contrarily, in the hot southern states of US southern live oak (*Quercus virginiana*) was planted before the Civil War (1861–5) to make cool avenues.

To see the world's longest tree avenue (as decreed by the *Guinness Book of World Records*) you need to visit the Tochigi Prefecture of

Avenues made of single species, like these beech trees, look impressive but can be decimated by the arrival of new pests and diseases

Japan, where Japanese cedar (*Cryptomeria japonica*) stretches for 22 miles. These giants (average height 27 m / 88½ ft) were planted in the years 1628–48 and converge on Imaichi City.

Planting monumental avenues of huge trees is rarely done now, because of a number of problems. Where trees are planted on both

Planting an avenue

Points to consider:

* The size of your route – how wide and tall will it be when the trees have matured?
* The size of tree you need and want – can you maintain it?
* Cropping – do you want to harvest fruit from your trees or will it drop on the ground and make a mess?
* Do you prefer evergreen, in leaf all year but casting shade on the ground and making underplanting harder, or deciduous, giving seasonal variation and easier underplanting but with leaf fall to manage each autumn?
* The characteristics of your soil and climate – which trees will thrive?
* Pest and disease resistance – have you researched the latest problems that might attack trees in your region?
* One tree type or diversity – do you want a grand entrance that could decline all at once in the future or a more informal look, with good provision for a range of wildlife?
* What about tree spacing and consequently the number of trees you will need to buy?
* And finally, can you afford to maintain your avenue using the professional tree care desirable for large and mature trees?

sides of a highway, they restrict the width of potential new transport schemes, which has led to political conflict. New tree pests and diseases have struck down mass plantings of elm, ash, horse chestnut and oaks in the UK. By planting several different (diverse) species together, diseases spread more slowly, without the ability to hop from tree to tree. The devastation of a whole avenue is avoided.

Consequently, the Sauchiehall Avenue project in Glasgow features a selection of different species, including *Acer campestre* 'William Caldwell', *Carpinus betulus* 'Fastigiata', *Acer platanoides* 'Deborah' and *Ginkgo biloba*. These are upright, disease-resistant and tolerant of pollution.

But there are locations where the majesty of a single-species tree-lined avenue is still desirable. In Cleveland, US, a different approach was to plant each street with its own species to replace hundreds of elms lost to Dutch elm disease. Maintenance of each street is managed efficiently, with its trees developing and growing in a similar way. Also, it helps residents to navigate easily around the city while enjoying the uniqueness of their own neighbourhoods. No new disease is likely to destroy more than one or two species at a time, so most of the city's trees are safe.

ACER PLATANOIDES   ACER CAMPESTRE

Build homes

It's estimated that 53 tonnes of CO_2e (carbon dioxide equivalent of greenhouse gases) are emitted when a four-bedroom detached house is built of bricks and mortar, but there are many plant-based materials that are suitable for building with and far more environmentally friendly. Sometimes these are combined with inorganic materials to give them integrity, but not always, for example in the case of all-timber homes.

It's always best to use locally grown resources, to save on transport costs, and this can also help homes look right in their surroundings. The plant materials used for construction store the carbon absorbed by the plants during their period of growth, and this remains stored until the end of the home's life. Building with plants also reduces our depletion of non-renewable resources.

Hempcrete is a new building material made from the refuse created in the processing of flax, hemp or jute, known as 'shives'. This refuse is the part of the plant stalk that is not fibrous and mainly consists of cellulose. One cubic metre of hempcrete walling has been calculated to lock up over 100 kg (220 lbs) of carbon.

Another composite is cob, combining straw with earth. In modern times it was replaced by more sophisticated concrete, only to be reintroduced for eco-friendly homes. It's estimated that at least one third of the world's population lives in earth constructions. The purpose of the straw is to provide structural reinforcement and moisture distribution. It helps the cob to dry out quickly. Straw is a waste product from many types of crops.

Bamboo culms are strong, light-weight and elastic and their rapid growth makes them a reliable and sustainable building material

Straw bales are also used to make walls, not mixed with mud, but tied together by pins or mesh and rendered for waterproofing.

Bamboo stems (culms) make both the material from which homes are constructed (for example, nipa huts in the Philippines) and the scaffolding to build homes. In Hong Kong, bamboo is used as the scaffolding for builders repairing skyscrapers! The fibres inside bamboo culms, running axially (along their length) give it a higher tensile strength than steel. It's also lightweight and elastic.

In wooded regions, timber is the material most suitable for sustainable homes. We're familiar with low-rise wooden cabins, but the tallest existing wooden building is The Sanctuary of Truth in Thailand, a temple 105 m (344 ft) tall. The W350 project in Japan proposes a 70-storey skyscraper made of 90 per cent wood in a heavily forested area, to be completed by 2041. One aim for this building is to regenerate Japanese cedar and cypress forests planted after the Second World War by using the mature trees that are no longer growing quickly (and so capturing carbon) before replanting the forests.

While building a sustainable home from scratch is an activity for expert builders and architects, keen gardeners can grow a living roof, made of live plants, with just a little planning and advice. The easiest ones to grow are self-sustaining and planted with drought-tolerant sedum and sempervivum succulents, such as *Sedum acre*, *S. album* and *S. rupestre.*

Living roofs have many benefits beyond aesthetic ones, including providing habitats for wildlife, improving urban air quality, cooling cities, mediating heavy rain, insulating homes and reducing noise. And many gardeners who can't create a living roof over their homes make mini ones to cover sheds or bins.

Growing a living roof

* Check the structure beneath the roof will hold its weight (you need 5–15 cm (2–6 inches) of growing medium for the plants to root into)
* Consider the access that will be needed by those planting and maintaining the roof
* Insulate and waterproof the roof, then apply a root barrier membrane
* Allow for drainage, with a drainage layer that cannot become clogged
* Choose and lay the right growing medium – it should contain lightweight materials, such as perlite
* For a self-sustaining roof, plant with sedum grown in geotextile pockets
* Be prepared to pull out weeds, like buddleia seedlings.

Drought-tolerant plants, such as succulents, help living roofs stay green without irrigation

Mitigate pollution

Have you ever seen a filthy hedge next to a main road and wondered how it got to be so dirty? That grime is caused by a combination of fumes, particulates from brakes and tyres as well as dust kicked up by passing vehicles, all of which is harmful to us, especially when we breathe it in. In fact, poor air quality is the largest environmental risk to UK public health. The good news is that on the other side of that hedge the air is cleaner because harmful particles have been trapped in the foliage.

Scientists are keen to quantify the extent to which plants can reduce harmful pollution. As far back as the 1980s, the space agency NASA carried out the Clean Air Study, looking at the ability of houseplants to remove volatile organic pollutants from the air. *Chrysanthemum morifolium*, *Rhapis excelsa*, *Chamaedorea elegans* and *Spathiphyllum* 'Mauna Loa' all scored highly, and in 2004 it was proved that benzene is removed from the air both by certain potted plants and the microorganisms in their compost.

But translating that data into practical use has proved difficult. To get the benefits it's thought you need an impractical density of plants for most interiors.

With hedges and road pollution, the case is stronger. The latest research claims that in seven days, one metre of well-managed hedge can mop up the same amount of pollution a car emits over a 500-mile drive. The best hedge species are those with hairy, rough and complicated leaves. Both *Cotoneaster franchetii* and *Thuja plicata* did well in the trial.

So why the focus on hedges over trees? Because road pollution is more prevalent low down – at a child's rather than an adult's head height. With trees the canopies are usually too high to catch it, while herbaceous perennials and annuals lack the bushiness and year-round foliage necessary to help. Hedges grown for this purpose should be bushy from their base upwards and at least 1.5 m (5 ft) wide. Our hedgerows provide many 'ecoservices', and this one is vital.

Scientists found the hairy leaves of *Cotoneaster franchetii* were particularly effective at soaking up roadside pollution

Although *Thuja plicata* grows into a big tree, with its intricate foliage it also makes a handsome hedge that protects gardens from vehicle pollution

Solve crimes

This is a gruesome chapter, featuring humans at their worst, committing abominable crimes, but also humans at their best, working out what happened with the help of plants.

The case often cited as the first to use forensic botany (that is, analysing plants to solve legal cases) concerns the child of a famous American aviator, Charles Lindbergh. In 1932 the child was kidnapped and then murdered. Arthur Koehler, a wood expert from the US Forest Service, analysed a wooden ladder used in the kidnap and discovered that its materials matched flooring in the suspect's attic. The evidence led to the suspect's conviction for kidnapping the boy for a ransom and murdering him. Both investigator and murderer would have had an intimate knowledge of wood anatomy as the killer was a carpenter.

Botanists at Kew can identify species of wood from small flakes examined under a microscope to reveal four different cell types – vessels, fibres, 'axial parenchyma' and 'rays'. This analysis is often used to tell whether smuggled objects brought into Heathrow are made from endangered hardwoods.

Sixty years after the Lindbergh case, DNA testing of plants was added to the detective's armoury. The RAPD (Random Amplified Polymorphic DNA) technique can match samples from the same individual plant. A scrape on a palo verde (*Parkinsonia florida*) tree at a murder scene led a sleuth at Arizona University to sample the DNA of a seed pod found in the back of the suspect's truck. When DNA

samples were taken from several different palo verde trees, the seed pod's DNA only matched one – the one with the scrape in it. The truck driver was convicted.

Forensic botanists will examine barbs from seeds left in clothing, plant material in a corpse's stomach, pollen on clothing and chemicals from plant-based poisons. The strong compounds from which plants are made, including cellulose, pectin and sporopollenin, are traceable long after a crime. Investigators can locate a buried body by spotting extra green vegetation close to it because the decay of a corpse creates a 'necrobiome' (a collection of organisms living on it). Large amounts of nitrogen released by a body during this process feed the plants nearby, resulting in lush growth with sinister connotations for murder detectives.

Under a microscope, the cross section of a plant stem reveals a unique pattern of vessels and fibres from which a species can be identified

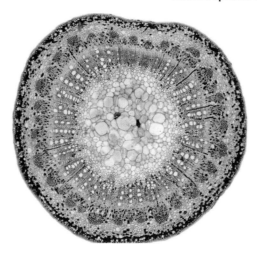

Increase property values

The evidence that proximity to green spaces increases the value of housing is now crystal clear, with plants playing a crucial part in those places, giving opportunities to engage with nature as well as relief from stress and a boost to happiness and wellbeing (see Stimulate and soothe mind and body, page 102). Even before the COVID pandemic confirmed how important parks are to the nation's psyche, the UK's Office for National Statistics (ONS) carried out research that revealed urban green spaces raised nearby house prices by an average of £2,500 per property.

The ONS analysed more than one million sales on the popular property website Zoopla between 2009 and 2016, and cleverly matched them with data from Britain's national mapping agency, Ordnance Survey (OS), which indicated where green spaces are located. These green spaces consist of public parks and gardens, play spaces, playing fields, sports facilities, golf courses, allotments, community growing areas, religious grounds and cemeteries.

While residents gained benefits from dog walking, sport and exercise in green spaces, the ONS also assessed their aesthetic importance by searching for property descriptions that included 'outlooks' and 'views' of green spaces and water. These attractive prospects boosted the home prices by an extra 1.8 per cent (£4,600 on average).

This methodical research aligns modern society's values and preferences with those of the nineteenth century, when Frederick

Law Olmsted relied on 'the proximate principle' to demonstrate that the impact of Central Park in New York on adjacent property values amounted to the park making a 'profit'. In 2017 a review of 30 studies on the usefulness of the proximate principle confirmed that it still holds true, and that 'passive' parks (landscaped areas without sports fields and facilities) increase property values more than active ones.

Clearly, plants and gardens are crucial to the price uplift. Wealthy owners of park-side homes rejoice in these statistics, but fortunately urban planners are aware of them too as they consider the welfare of city populations, rich and poor.

Research on the kind of domestic garden house buyers are attracted to shows most favour tidy gardens that are easy to maintain, so if you're revamping a garden ready to sell your home, then the kind of plants that are sometimes called 'supermarket car park' plants should help you create the right effect with little effort on your part.

EASY-MAINTENANCE SHRUBS

- *Cotinus*
- *Hebe*
- Holly
- Snowberry
- *Brachyglottis*
- Oregon grape
- *Cornus alba*
- *Lonicera nitida*
- Heathers
- Groundcover roses
- *Forsythia*
- *Spiraea*
- Ninebark

Fence a boundary

'Good fences make good neighbours' Robert Frost, *Mending Wall*

Fences have multiple uses as barriers, for seclusion or to enhance a place, and while metal will often do the job, wood is a renewable resource that can be worked by anyone with basic skills. From English post-and-rail fences to *zariba* thorn fences in Sudan, these boundaries are often culturally specific and can indicate the regional location of a landscape as effectively as growing plants.

Because wood rots when exposed to air and moisture at the same time, no purely wooden fence will last forever, as it will eventually give way at the soil line. Modern fence posts are usually embedded in spiked steel supports or concrete, and formal fences will need to be made from hardwood or pressure-treated softwood and maintained with preservatives.

MAKE A ROUNDPOLE FENCE

- Drive pairs of 1.5–2 m (5–6½ ft) uprights into the ground along the fenceline
- Using ties made from young saplings, loosely tie each pair together around 20 cm (7¾ in) from the ground
- Lay down longer, round branches as slanting horizontals with one end resting on the ground and the other on a tie
- Repeat with more slanting horizontals and ties until you reach the tops of the uprights.

Fences made from simple stakes, with branches woven between them, are attractive, easy to make and sustainable

Rustic fences, however, can be purely plant-based. These are the easiest to construct from scratch and can be formed entirely from natural materials, even leaving bark on to save time, give a distinctive look or provide a habitat for garden wildlife. The timber may be waney-edged (that is, unsawn and retaining its natural shape), saving the resources that are put into creating timber planks.

There are several traditional designs, of which my favourite is the roundpole style typically found in Scandinavia and Estonia, where it's made from spruce or juniper. This method is known to date back at least to the Iron Age.

Other informal boundaries are woven hurdles of willow, or oak strips, split bamboo screens and barriers of brushwood simply stacked between pairs of poles. Brushwood fences provide habitats and resources for birds, amphibians, insects and many other creatures.

Improve office productivity

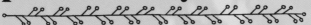

The word 'productivity' is harsh and dry, and makes me think of people in 1980s power suits trying to squeeze the last drop of effort out of overstretched office workers. But we can all benefit from being more productive if it means we make better use of our working hours – helping us feel more relaxed at work and avoid anxiety during leisure time. Of course, not everyone's working day involves an office, but the pervasive spread of computers seems to have made deskwork more, rather than less common. It's growing plants indoors, near desks and inside offices that I want to focus on here, as workers who are based outdoors are more likely to have plants nearby already.

A number of benefits from houseplants have been identified by human resources departments. Office plants invoke feelings of calm and peacefulness, improve mood, promote creativity, memory and attention span and reduce sick-day figures, according to studies. By absorbing sound (turn to page 44 for more on this) they can reduce distractions and aid concentration. Researchers at London South Bank University concluded that positioning larger plant pots in multiple locations around a room provided the best sound buffering.

Plants' ability to reduce air pollution (which I cover in more detail on page 54 is particularly pertinent in a shared office situation, where wipe-clean, synthetic furnishings give off chemicals such as benzene, trichloroethylene and formaldehyde. As well as these particularly harmful gases, plants reduce carbon dioxide and nitrogen dioxide (NO_2) levels.

One study found that indoor plants can help reduce CO_2 levels by about 10 per cent in air-conditioned offices, and by about 25 per cent in buildings without air conditioning. Palms were found to reduce CO_2 levels the most. Because raised CO_2 levels make us drowsy, it's thought office plants could counter the mid-afternoon slump as effectively as caffeine! Ongoing research by the Royal Horticultural Society and the University of Birmingham is investigating how indoor plants can reduce harm from NO_2.

The acme of this approach is Amazon's HQ in Seattle, where three glass domes containing an 'indoor rainforest' have been constructed for workers to enjoy in an effort to encourage creativity and invention. Workers can meet colleagues, bring recruits (research shows plants make work spaces more attractive to job applicants) and hold team meetings there to enjoy the recreation of a cloud forest, with a 20 m (65½ ft)-high tropical fig and rare begonias from the Philippines.

Amazon's real estate chief, John Schoettler, was inspired by Kew Gardens, 'We thought: What would Kew Gardens be like if it were being built today?' Crucially, this space smells like a lush forest. Amazon is not the only big tech company looking to bring nature to work. Google, Apple and Meta (formerly Facebook) have nature-inspired projects of their own.

Regardless of the care regime, plants in offices need to be robust, so go for the tough ones that survive infrequent watering and dry air, grow slowly and don't need pruning and primping. Avoid spiky ones too!

The ultimate list of office unkillables includes the following: *Epipremnum aureum, Sansevieria cylindrica* and *S. trifasciata, Aspidistra elatior, Monstera deliciosa, Zamioculcas zamiifolia, Howea forsteriana* and *Dracaena fragrans.*

ACTION PLAN FOR YOUR PLANT-BASED OFFICE MAKEOVER

ASSESS THE ROOM CONCERNED
* Take day and night temperatures
* Find out which direction the windows face
* Consider humidity – the air in most offices is too dry for many houseplants.

CONSIDER HOLIDAY TIMES WHEN THE OFFICE IS EMPTY, LIKE THE CHRISTMAS BREAK
* If the heating will be off, plan to rehome or replace tropical plants.

DECIDE WHICH PLANT TYPES YOU NEED – MATCH THEM TO THE OFFICE ENVIRONMENT
* Work out how to improve the plants' environment – do you need trays of wet pebbles to improve humidity or window frosting to reduce harsh sun?

MAKE A PLANT MAINTENANCE PLAN
* Book a plant maintenance company to visit your office regularly, or
* Find the office plant-mad person and make plant care part of their role, or
* Devise a rota so lots of workers share the maintenance.
* Don't forget to arrange training for those who take part.

North-facing window light is not too strong for house plants

Care rota

Humidity from wet pebbles

Room thermometer

West-facing window might give too much afternoon sunshine for plants

Storage space for care equipment

Clothe and comfort

Dye cloth

If you're hoping for a rainbow of bright zingy colours, then naturally dyed cloth is not for you. But if you appreciate a range of ochre, dusky pink, biscuit, oat, umber, sage and chartreuse, you'll be delighted by the hues you can create from plants found in ordinary gardens. Perhaps more unusual is blue from woad (*Isatis tinctoria*), but even this striking plant is easy to grow from seeds that can be bought from seed merchants. These natural colours sing off each other, meaning they're at their best in a quilt or Fair-Isle scarf combining several dye batches.

Picture a pear tree, flanked by juniper, elder and cotinus. A hop rambles over the shrubs and at their feet are sorrel, tansy and marigolds. All of these are plants that might be grown in a dyer's garden. Different plant parts are used and must be harvested at the right time: leaves when fresh and young, flowers just coming out and fruits in season.

Some dyers use a 'mordant' to fix the dye so it doesn't run. These are often poisonous chemicals such as chrome and copper and they're specific to the dye plants. You prepare the fabric by boiling it in the mordant before immersing it in the dye mixture. Natural fabrics take natural dyes best, with wool being the easiest to use.

Wear gloves when using mordants to protect yourself from the poisons and put on protective clothing whenever dyeing, as well as doing it somewhere you can make a big mess – plant dyes are powerful and prone to colouring everything they touch.

Dyeing is a lengthy process. Here's an example of the steps needed to make green wool yarn from elder leaves:

- Gather your equipment – undyed yarn, large stainless-steel boiling vessels, tongs, measuring jug, protective clothing, pestle and mortar, thermometer, soft water and scales–
- Weigh, then prepare the yarn by washing in soap flakes to remove grease. Rinse several times and add vinegar to the final rinse
- Make up the mordant: elder needs alum. Dissolve the alum in hot water then add the yarn and simmer for an hour. Remove the yarn and rinse
- Pick your elder leaves – you'll need roughly the same weight of leaves to dry yarn
- Grind the leaves in the mortar and secure in a muslin bag
- Leave the muslin bag of crushed leaves soaking in tepid water overnight
- In the morning, bring to the boil, simmer for 1–3 hours, then allow to cool to hand temperature
- Add the yarn and bring slowly to the boil again. Simmer for an hour, then take the pan off the heat
- Leave the yarn in the liquid until it's cold
- Remove the yarn with the tongs and rinse until no colour comes out, then dry.

USEFUL DYEING PLANTS

- Comfrey
- Elder
- Juniper
- Nettle
- Onions
- Sorrel
- Tansy
- Walnut

Make bed clothes

C STANDS FOR COTTON
Its beautiful bolls,
And bales of rich value, the Master controls.
Of "mud-stills" he prates, and would haughtily bring
The world to acknowledge that "Cotton is King."
–*The Gospel of Slavery*, **by 'Iron Gray' [Abel C. Thomas], 1864**

I'm not going to suggest you spin and weave your own cotton sheets –
it's taken thousands of years to develop the technology to turn cotton
goods from luxuries for the nobility into everyday items for ordinary
people, and it's a very complicated process. Dyeing your bedding
with plants from your garden (turn to page 68) would be much easier.
To grow cotton plants – various species of *Gossypium* – you need a
steady temperature of around 20°C (68°F) with high humidity for five
months, so it's not a temperate crop.

Cotton, from the finest Egyptian or pima types to the more basic
'upland' type, is the most popular global bedding material. This
renewable resource, when pure and not combined with synthetic
fibres, is biodegradable at the end of its life and won't shed harmful
microfibres in the wash. It's breathable, light, soft and can be
washed at a high heat, making it hygienic and therefore ideal for
hospital sheets.

But while you can dispose of a cotton pillowcase on your compost
heap, the making of it is likely to have used huge amounts of
chemicals and water. According to the World Wide Fund for Nature it
takes 20,000 litres of water to produce just 1kg of cotton on average.

This quantity can be reduced with carefully planned irrigation and land use. The Sustainable Trade Initiative states that no commodity is as polluting as cotton, with about 10 per cent of all agricultural chemicals employed globally being used in the cotton sector.

When the crop is ready for harvest (around 55–80 days after sowing), it looks like a field of cotton wool. The fluffy white 'bolls' are a protective case around the seeds and are the raw ingredient for the fibres. Before harvesting, the plants are defoliated, a job still usually done by people before harvesters (usually machines) collect the bolls, removing large contaminants and making bales. The machine that separates the oily seeds from the bolls is called a gin, and its development in America is thought to have been a cause of the American Civil War, when easy cleaning increased the demand for cotton grown by slaves in the South and woven by the textile industry in the North.

Gossypium is one of the most important plants in the world, now and historically. But its products can come at a high environmental and social price, so please value your sheets and pillowcases, and dispose of worn-out bedding thoughtfully by reusing, recycling or composting it.

GOSSYPIUM HERBACEUM

Create a linen scarf

Linen cloth, made from the flax plant (*Linum usitatissimum*), is harder to manufacture commercially than cotton, but easier for temperate home gardeners to make from scratch. Use of this annual plant dates back at least 30,000 years.

Flax is ready for harvest for cloth in around 100 days when the stem turns yellow. To grow for the oily seeds (linseed) you leave it longer, to maximise the crop. The plant prefers rich, fertile soils. Pulling the plant by hand retains the longer fibres (up to 1 m) for which linen is valued.

Run the pulled plants through a riddle to get rid of seed pods. Ret them in water, that is, allow them to partially rot. This breaks down the pectin in the stems, to separate the phloem structures you want to keep from the xylem structures you want to discard. Belgium and Ireland are traditional centres of linen production due to having good conditions for retting.

Next, crush the stems in a flax break. This releases the long, useful, hair-like outside (bast) fibres. Then scrape them on a scutching board, to improve the quality of the fibres, and brush them out through a heckle comb made of nails. The comb removes the short fibres you don't want to spin with, although these can be used for other things, like pillow stuffing. Finally, you're ready for spinning! After spinning, it's easiest to weave rather than knit linen thread as it has no elasticity.

This may seem like a long process, but if you do it you'll be in good company, as the finest linens are still manufactured by hand! Linen is a contender for the prize for the most eco-friendly fibre, along with hemp. Hemp, from *Cannabis sativa*, also provides bast fibres for weaving. It's simpler and more efficient to produce than cotton, releasing fewer toxic substances, although it requires more nitrogen to grow. There are cultural reasons why it's more expensive, relating to the dominance of the cotton trade and the illegal use of cannabis as a recreational drug.

Look out for beautiful fields of flax if you're driving through Britain in summertime. En masse, the flowers look like an azure haze

Design a floral fabric

Look around any fabric shop or website and you'll notice that the number one inspiration is flowers. Their intricate but organic shapes intrigue and delight us, meaning that designers have copied, manipulated and exaggerated their forms for centuries. From William Morris, through Zika Ascher, Mary White and Lucienne Day, to Cath Kidston and Orla Kiely, blooming nature has provided the motifs for famous patterns.

Examining one of the complicated fabrics of these designers you might wonder how they've managed to repeat the pattern without any joins showing. In fact, it's really easy to achieve this so I'm going to describe the simplest method, but you can adapt it once you've mastered the basics.

This photo of rudbeckias and selinum would make a good starting point for a floral repeat pattern

How to repeat a pattern

1 Choose a photograph of a group of flowers, such as the one below left. Both flowers are easy subjects for stylised drawings, and growing together they already form a pattern that reminds me of a fabric.

2 Gather a piece of paper (square or rectangular), a pencil, rubber (eraser), scissors and sticky tape.

3 Draw a design in the middle of your paper. Keep it away from the edges.

4 Cut your paper in half from top to bottom. Then tape (on the back) the sides together so the old middle of the design is now on the sides.

5 Cut your paper in half from side to side and tape (on the back) the old top and old bottom together so the old middle is now at the top and bottom.

6 Draw more flowers and foliage into the middle of the paper. Your repeating design is now complete.

7 Use your design to make a repeating pattern by transferring it onto lino, cutting into the paper to form a stencil or scanning it into a computer. This is where the fun starts! Why not print onto your homemade paper from page 159 or your linen fabric from page 72?

Repeat patterns use one block placed on the fabric several times, but you can't see the join

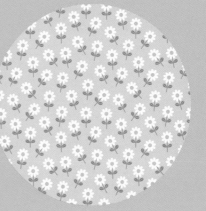

Green up a garden

Improve garden soil

A vital component of garden soil is organic matter, much of which originates in plants themselves. Part of the reason why deserts are inhospitable to plants and other organisms is that their soil lacks organic matter and is mostly composed of mineral particles, like sand. This can be a vicious circle, in which fewer plants growing results in less soil organic matter, which makes it harder for plants to grow and consequently the organic soil content reduces further.

You might be wondering why plants need organic matter in soil. A healthy soil consists of humus (the organic component of soil, formed by the decomposition of plant material by soil organisms) and inert mineral particles. These are held together by fungal strands and organic glues, made by bacteria, and the same fungi and bacteria eat organic matter.

Earthworms also eat it and as they burrow through the soil, they make channels, allowing water and air to penetrate the soil. As the

humus gradually breaks down it also releases plant nutrients such as nitrogen. Through this process soil gains the three things plants need from it – water, air and nutrients.

Plants naturally provide this organic matter for the soil as they shed parts they don't need – dead leaves, outer bark, the fleshy parts of fruit. In a forest you can observe this process happening when leaves fall to the ground in autumn and break down over the following months until they disappear from sight.

Interaction between this 'leafmould' and soil organisms results in the organic component of the woodland soil, essential for the life of the trees and completely independent of human intervention. The system feeds itself.

In our gardens we can mimic this process by adding organic matter, such as home-made garden compost, as a mulch to any bare soil in our borders. As the soil organisms process it the border soil will improve with no further effort from us. Turn to the next page to find out how to make your own compost.

Natural leaf litter (consisting of fallen autumn leaves) helps both these crocus (left) and birch trees (right) to thrive. Note how the crocus buds easily push up through the layer of leaves

Make compost

We expect our garden plants to be more productive than wild plants, with bigger flowers, more fruits, greener leaves and longer periods looking good. This means improving our soil, and making garden compost is the cheapest and most sustainable way to do that. Prunings, parts of vegetable plants we don't eat and flower deadheads are ingredients for the compost heap, which will break down into a free and local source of mulch and soil improver.

Well-rotted compost like this can simply be placed on top of soil to be incorporated by organisms such as earthworms

Add spent flowers, vegetable haulms, grass clippings and fallen leaves to your compost bin or heap

Compost-making basics

- Turning compost makes it mature faster
- Garden compost needs both 'brown' and 'green' ingredients
- Provide browns from woody shrub prunings and greens from lawn mowings
- Mix up your greens and browns
- Chop up browns before adding them so they break down faster
- Don't add weed seeds or roots unless you've killed them by soaking (seeds) or drying (roots)

- Don't add meat or cooked veg as they attract rats
- The bigger the heap the faster compost will be made
- Don't worry if your heap is small, it's still useful
- Hot composters and wormeries work well for enthusiasts with only small gardens
- A line of bins or bays works best, moving the compost from raw ingredients to finished product bin/bay as it matures
- Mature compost smells good
- Spread it over any bare soil, like a tree shedding leaves, and let the soil fauna mix it in

Collect your flower deadheads together to become compost heap ingredients and eventually part of the soil for growing new blooms

Pretty *Pontaderia cordata* flowers
from July to September and provides
shade to deter algae growth

Clean/maintain garden pond water

A garden pond with a diversity of wildlife and a healthy ecosystem has clear water and smells good. There will be a balance of life coexisting. Although it will contain a small amount of algae, the algae won't be dominant so there will be no floating scum, pea-soup effect or obvious blanket weed.

Algae not only look unattractive in garden ponds, they also deoxygenate the water, making it an unsuitable habitat for other life. Algae thrive where ponds contain excessive nutrients, such as nitrogen, due to contamination with fertilisers, the breakdown of decaying organic matter or fish poo.

One of the simplest and most effective ways to maintain the balance of life in a garden pond is to add 'oxygenating' plants to it. In fact all plants are oxygenators as when they photosynthesise they take in carbon dioxide and give out oxygen. But garden pond enthusiasts refer to some types of pond plant as **oxygenators**, and these are the ones that are generally fully submerged in the pond. By living underwater, their oxygen is released into the water where it benefits other pondlife. Pond creatures also enjoy hiding among the foliage.

Because oxygenating plants take up nutrients as they grow, they compete with algae, making it harder for algae to thrive. To maintain the balance, it's important to thin out the oxygenating plants occasionally so that they themselves do not choke the pond in a similar way to algae. The spacing recommendation for adding them is just four bunches to a square metre, each bunch having three to four stems.

Floating plants also deter algae, which thrive in sunny conditions and grow less well in water shaded by plants like lily pads. Ideally about 50 per cent of the pond surface should be clear of vegetation, so don't let plants take over the pond.

For larger ponds, adding a bale of barley straw may reduce algal problems. It's placed in the water in spring and removed when it turns black around six months later. It's thought that while this plant decomposes in the pond it releases chemicals that deter algal growth.

With the right plants no pond pump is necessary to keep a garden pond clean – in fact, natural swimming pools can be constructed with submerged oxygenators and planted gravel on the margins. Garden ponds are vital for pond wildlife now as 70 per cent of wild ponds have been lost from the UK countryside.

OXYGENATING PLANTS

- *Callitriche autumnalis* – water starwort
- *Ceratophyllum demersum* – hornwort
- *Fontinalis antipyretica* – willow moss
- *Hottonia palustris* – water violet
- *Myriophyllum spicatum* – spiked water milfoil
- *Potamogeton densus crispus* – frog's lettuce

Support garden wildlife

One of the most satisfying things you can do with plants is to support your garden's wildlife. As we enter the Anthropocene (the era in which human activity is having a significant effect on our planet's climate and ecosystems) our responsibility for the creatures around us grows ever stronger. Our gardens take up areas of land that once were wild, so they are still populated by a wide array of wild fauna, but with thoughtful gardening we can also make a home for creatures that have nowhere else to live.

While the design of our gardens can have a significant impact on wildlife, what we plant there is even more important. One particular group, the pollinators, provides services for us that we couldn't survive without. Pollinators increase the output of 87 per cent of crops worldwide. Look for lists of flowers that attract bees or butterflies in your area and grow those to encourage pollinators in your garden. Monitor plants that insects visit and grow more of them.

If you don't have a garden to support wildlife, a window box or container on a city balcony can be an important 'nectar bar' for travelling pollinators looking for an energy boost. Plant up with attention-grabbers like alliums, geum and **achillea,** pop in some trailing thyme and nasturtiums to cover the edges, then fill in the gaps with astrantia, nepeta and oregano. All of these have flowers that insects love and you'll enjoy watching their visits.

Common garden wildlife in the UK includes (top row) the holly blue and small tortoiseshell butterflies, robin and bumblebee. Hawthorn fruits and the nectar of verbascum, poppies and allium sustain wildlife

Flowering trees and shrubs can make a great pollinator restaurant too. Flowers that have a long blooming period can provide the best service, and ironically some flowers that are bred to be sterile are great for this. Geranium 'Rozanne', for example, never goes to seed so keeps flowering all summer and into autumn and consequently makes nectar for months.

Don't forget night-flying moths, which tend to prefer heavily scented, pale, night-blooming flowers like *Nicotiana*.

Butterflies and bees are attracted by colour, but you might be surprised by the flowers they flock to – they see colour very differently from us and can perceive patterns on petals, invisible to our eyes. For the greatest benefit, grow lots of differently shaped flowers, but don't give space to blousy double flowers as they often have no pollen or nectar available.

Your butterfly and moth populations will thank you for providing larval food plants. These are the ones that caterpillars eat, and they're different from the striking blooms the adult insects visit. Many of them are plants you'd consider weeds – stinging nettles, brambles, clovers and grasses are essential for certain moth and butterfly species. Keen wildlife gardeners will tolerate or even deliberately grow these.

Do grow some winter-flowering plants too, for insects like queen bumblebees, which need them to survive the colder months.

To help birds, give them the shelter of trees, shrubs and climbers, and forage in the form of berries and seeds. Grow a hedge instead of erecting a fence. Many garden birds eat insects in summer and seeds in winter. By looking after your pollinators, you will be feeding the

birds too. It may sound gruesome, but the birds need this food, which is part of the web of life.

Think about surfaces – mown lawns are not very diverse, but they're better for wildlife than barren concrete (see page 98). Lawns where wildflowers have been allowed to grow are better, and you could make a clover, thyme or camomile lawn. Letting a lawn grow long for part of the year and making a meadow is great, too.

Soil fauna is the garden wildlife scientists know the least about, but it is hugely important. You can look after your soil, increase its fertility and encourage soil dwellers by avoiding bare patches. Keep your garden planted and sow a green manure to dig in if you have nothing else. Plant matter (roots and decaying foliage) in the soil is eaten by the herbivores there, and they in turn are consumed. The soil is an essential part of your garden's ecosystem and cannot be underestimated.

It can be hard for pollinators to find pollen and nectar in the colder seasons, so help them out with winter-flowering plants like crocuses

Build a den

This activity amuses children, but plenty of youthful adults will enjoy it just as much! There are many benefits for children in den building, including learning about nature, teamwork, cooperation, independence and problem solving, along with all the benefits connected with being outdoors and imaginative play.

These of course need to be balanced against safety considerations to protect the children from sharp objects, allergens and other hazards. Pulling bracken isn't recommended in places where there are ticks, which can carry Lyme disease.

Charities such as the National Trust and Woodland Trust are keen for children to try den building, but they ask for the dens to be deconstructed at the end of the day. If you have space for a den in your garden it can continue to be developed indefinitely.

A scout-tent den can be free-standing or started by leaning sticks on a long, low branch

Cover the finished den with leaves to make it extra cosy

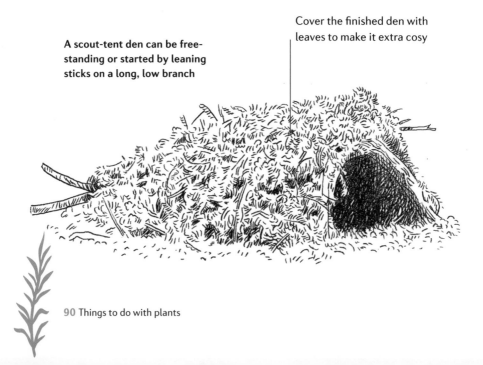

Start by finding the perfect spot. Choose somewhere with a comfy floor for sitting or lying in, not wet, bumpy or sloping. Long-term dens benefit from light and heat from south-facing doorways.

Collect long sticks, bendy sticks, big leaves or grass and a tying material. String is ok but you could try using vine stems for tying or

Fan long sticks out from a natural tree fork for a tepee-style den and tie them at the top if you want it to last

Weave bendy sticks in and out of the main framework

cordage you've made yourself (see page 145) if you want a completely natural den. Never cut branches if you're in a wild spot, but you might be able to use sticks from somebody else's dismantled den.

Two good den designs are scout-tent-shaped and tepee. For the first place a long stick between the low forks of two neighbouring trees. Prop sticks against it on both sides and weave the bendy sticks in and out of them before covering with leaves.

For a tepee, prop long straight sticks against a forked tree, fanning them out and then weave the bendy sticks in and out. With a tepee it's easy to leave a gap for a window. Cover the walls with leaves.

To finish off either den you can add mud on the outside and cosy leaves on the floor.

Look out for abandoned dens you can dismantle to start your own woodland hideaway

Create a bouquet

There's something special about a bouquet. It is not just a random selection of blooms plucked from a garden, but is designed, picked with care and arranged beautifully, to give to someone else.

For a perfect bouquet you need to consider harvesting, prepping, arranging, storing and presentation. This topic lends itself to an action plan, because following a few simple tips could turn your creation from a wilting washout to a glorious gift.

Roses, sweet peas and love-in-a-mist seedheads are mixed with heuchera and alchemilla flowers to make a very personal gift

BOUQUET ACTION PLAN

YOU WILL NEED:

- a bucket
- water
- secateurs or snips
- gloves
- wrapping paper, kitchen paper or similar

HARVESTING

- Wear gloves if you have sensitive skin. Plant sap can cause nasty allergic reactions

- Aim to pick first thing in the morning. Plants are at their most turgid then, having taken up water during the cool of the night

- Walk round the garden before you start picking to decide what to harvest

- Note the stems that are looking good that morning. Any flowers or leaves that are past their best will ruin your efforts so forget any ideas you had last night and really look at the condition of your plants today

- Decide on a scheme of colours, textures and shapes that set each other off. It's easiest to work with threes, fives and sevens to achieve a pleasing design, but pick a few more than you think you need

- Prepare a bucket of cool water in the garden before you pick and place each stem into it, up to its neck, as soon as you pick it

- Choose foliage as well as flowers, as it sets off flower colours better than anything else

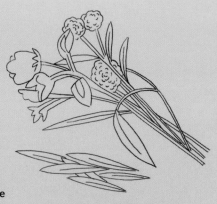

🌺 Most blooms are best picked before they're fully open

PREPPING

🌺 Keep the stems in water somewhere cool until you're ready to arrange them. Don't leave it for long!

🌺 Strip off the lower leaves from stems as any leaves below the water level in a vase will rot. Sear the base of euphorbias and poppies in boiling water.

ARRANGING

🌺 Lay out a square of wrapping paper and place three or five of your tallest stems across the diagonal

🌺 Add the other stems in threes, fives and sevens, progressing down by height

🌺 Wrap the base of your stems in moist paper then pull your wrapping paper around the stems

PRESENTATION

🌺 Hand over your bouquet as soon as possible after arranging it. This is crucial!

🌺 Advise your recipient on which stem bases to recut before putting it in a vase

🌺 Ask them to keep it in a cool spot, especially overnight, and keep the vase topped up with water

Control plant pests

It's not surprising that evolution has resulted in plants producing chemicals to control the pests attacking them and that some of these chemicals are harnessed by gardeners to direct them most effectively. Products made with plant-derived chemicals can be part of an organic regime, with less persistence and impact on the environment than purely synthetic chemicals. However, their action on the pests they control is still gruesome.

Pyrethrins attack the nervous systems of insects and act as a repellent for those they don't kill. They are biodegradable, decomposing on exposure to light. The insecticide (meaning, insect killer) is made from the seed cases that follow the pretty white daisy-shaped flowers of *Tanacetum cinerariifolium* (also called Dalmatian chrysanthemum).

It's known as a broad-spectrum insecticide because it kills a wide range of insects, including whitefly, caterpillars, aphids, leafhoppers, thrips, capsids, ants and beetles.

Plant oils, often made from rapeseed or sunflower seeds, are applied by gardeners to block the pores of the insects and mites they come into contact with. These creatures breathe through pores all over their bodies, called spiracles, and blocking them causes suffocation.

The chemical azadirachtin, produced in the oily seeds of the neem tree (*Azadirachta indica*), is likely to have evolved directly to stop feeding insects. It is also a broad-spectrum insecticide, which acts as an anti-feedant, repellent and egg-laying deterrent. The insects starve to death within a few days.

Cabbages and other brassicas contain glucosinolates, the chemicals responsible for the bitterness of Brussels sprouts that are attractive to some butterfly pests, like cabbage whites. However, they reduce certain soil pests, such as harmful nematodes. Consequently, green manures, grown from the hot mustard plant 'Caliente', are effective soil fumigants when dug into the fallow soil. This will also increase the humus content of the soil (see page 78).

These are only a few examples of plants used by gardeners to aid pest control. A more passive one is the carnivorous *Pinguicula* or butterwort, a sticky-leaved plant used by greenhouse gardeners at Kew. It traps small flies and gradually digests them using enzymes to break down their bodies, so their fluids can be absorbed into the plant through its cuticle.

The insecticide pyrethrin is made from the seedheads of Dalmatian chrysanthemum or *Tanacetum cinerariifolium*

Grow a landscaping surface

Pity the poor lawn-grass plant. It's walked, driven and sat on, cut on average 38 times a year (UK figures), rarely (if ever) watered and sometimes scraped with a scarifying rake or covered with soil as a 'top dressing'. In return it puts down roots so tough it survives weeks without rain, outcompetes most weeds, knits together companionably with its neighbours and looks lush for most of the year.

Numerous species of grass are used to make a lawn, and these have been bred further to tolerate more shade, be cut or walked on more often, or put up with heat and drought better. But all can survive frequent mowing and even thicken up as a result – throwing out tillers (side shoots) in response to the mower. Most other plants can't cope with mowing, unable to take the regular removal of their photosynthesising organs. Some other plants, which also have low growing points like grasses, can cope and these may be labelled lawn weeds if you're looking for a perfect sward.

Lawns are used as paths, car parks and sports pitches, situations where hard surfaces are alternatives. But those hard surfaces can't capture carbon the way a lawn can and often fail to let rainwater through, so they contribute to flooding after storms. It's easy to naturalise spring bulbs, such as snowdrops, crocus and narcissus, in informal lawns, but do wait for the leaves to die down naturally after flowering before you mow, as this allows the plants to sustain themselves by photosynthesis.

Letting lawns grow long, to become flowery meadows, has become fashionable and attracts a greater diversity of insects and other wildlife. But that isn't a good alternative when you need a practical surface or want a plain foil for other garden features. Formal lawns are not strictly monocultures as they're made up of different grass cultivars, mixed to suit the situation, but they can be environmentally damaging if synthetic pesticides, herbicides and fertilisers are used, and water is wasted on them, with powered machinery using energy unnecessarily. Cutting lawns will contribute to atmospheric carbon even more if the clippings are not composted responsibly.

ACTION PLAN FOR A SUSTAINABLE LAWN

- 🌺 Prepare the soil perfectly before seeding or laying your turf
- 🌺 Use a manual (push reel) mower
- 🌺 Cut very frequently without collecting the clippings so they are tiny and can be left to rot down and feed the soil
- 🌺 If collecting the clippings, compost them (see page 80)
- 🌺 Avoid watering after establishing – a temperate lawn should survive high summer looking brown, then green up again in autumn rain
- 🌺 Tolerate non-grass plants that merge well enough to look good with the grasses
- 🌺 Dig out plants that look out of place, such as thistles and plantains
- 🌺 Scarify and aerate each autumn to help the grasses thrive
- 🌺 Only use organic feeds and pest controls

Stimulate and soothe mind and body

Get to know them

My question was, 'Why are all the dandelions in my lawn so short that the lawnmower passes straight over them, when I know they can grow tall and lush?'

A few months into my first gardening course I learned the answer: they can produce seeds by 'apomixis', that is, without pollination, so offspring are genetically identical to their parent. Once a ground-hugging dandelion starts making clones of itself in a lawn the population of short individuals can grow easily because the mower passes over without snipping off the seedheads!

Gardening courses are just one way to find out more about plants. You can study botany, ecology, forestry or agriculture at university, join a botanical society, enlist at a high street florist or volunteer at your local public garden.

Teaching yourself is possible too, growing your own plants and using websites and books from your local library or borrowed from an institution like the UK's Royal Horticultural Society.

Plant books have their own special names: a Flora is a publication that attempts to cover all of the plants growing in a region, whatever its size, from tiny islands to huge countries. Field guides are much smaller books that can be carried 'into the field' and therefore focus on plants you're most likely to see. They usually have a 'key' to help you identify the plants you come across, following a series of steps. At the other end of the scale, a monograph is a guide to one small group of plants.

As you get deeper into your study, you'll need to know how plants are named. I'm not talking about common names like bluebell, but their more precise scientific names. A bluebell, for example, might be the climber *Sollya heterophylla*, the spring bulb *Hyacinthoides non-scripta* or the moorland perennial *Campanula rotundifolia*.

Scientific naming starts with the species level (*C. rotundifolia*), groups these into genera (*Campanula*), which in turn are grouped in families (Campanulaceae) and onwards into bigger and bigger groups of genetically related plants.

Species can also be broken down into smaller groups, such as cultivars, which are pleasing examples of that species, selected or bred for a purpose. And if that's raised any questions in your mind it might be time to enlist on a course…

Sollya heterophylla is one of many plants to bear the common name of bluebell

Investigate maths

If, like me, you first became interested in plants because you enjoyed pretty flowers and found it hard to understand the point of studying mathematics beyond what was necessary for everyday life, you might be tempted to skip this section. But please don't, because when you find out how maths and plants relate to each other you'll have a new appreciation for the beauty of maths as well as blooms.

Outdoor classrooms have become common. By taking children out into the school garden, teachers give them fresh air, vitamin D and inspiration to fire up their brains, ready to take in their lessons better than before. Gardens are now used as a learning resource in subjects ranging from art to maths. It's now easy to find garden-maths teaching ideas online for children, but there are many botanical-mathematical concepts that will delight even maths-phobic adults, too.

We use maths to understand botany, for example in ecological surveys, where sampling techniques estimate populations of wild plants too large to count. You place a series of quadrats (small squares) at random, counting all the individuals within each square, then calculate how many individuals there must be in the whole area. Sampling techniques are employed to work out how many leaves there are on a tree, too, working from individual branches.

Another straightforward technique for assessing populations is the bell curve. How tall are all the individuals of one species? Plot their heights across a graph and chances are you'll create a bell-shaped curve, with most (ordinary) individuals clustered in the centre of the bell and a few (extraordinary) tall or short individuals spreading out on either side.

A slightly more complicated version of this involves plant growth and the way branches are distributed in space. Generally, growth is densest near the plant's centre (like an oak branch coming away from the trunk) and gets less dense further away (the twigs). When the growth of many different plant species is plotted on a graph,

Look out for spirals following the Fibonacci sequence in nature, as in the flowerheads of *Scilla peruviana* (left) or annual sunflowers (right)

a bell curve is the result, and researchers have found the same mathematical properties at work in brain neurons!

The most mind-blowing plant–maths relationship is found in too many plant types to list, some examples being pineapple fruits, Brussels sprouts, pinecones and sunflowers. It concerns phyllotaxis, the arrangement of plant parts around an axis. Plant leaves and flowers arranged in spirals hold together in a strong unit, without gaps, like a snail shell.

Amazingly, this kind of spiral is represented by a neat mathematic sequence, discovered by a thirteenth-century Italian mathematician, Fibonacci. In the sequence, each number is the sum of the two

preceding ones: 0, 1, 1, 2, 3, 5, 8, 13, 21, 34 etc. To visualise this, it's easiest to plot it on graph paper, spiralling from the centre. To see it at work in a sunflower head or pinecone you need to think of several spirals next to each other, all coming from the centre.

Sometimes it seems like we're not the only ones doing calculations. Scientists at the John Innes Centre realised that plants were able to use their starch reserves overnight at a constant rate, so that they run out at dawn, when they can start photosynthesis (and making more starch) again. Could plants be doing their own maths?

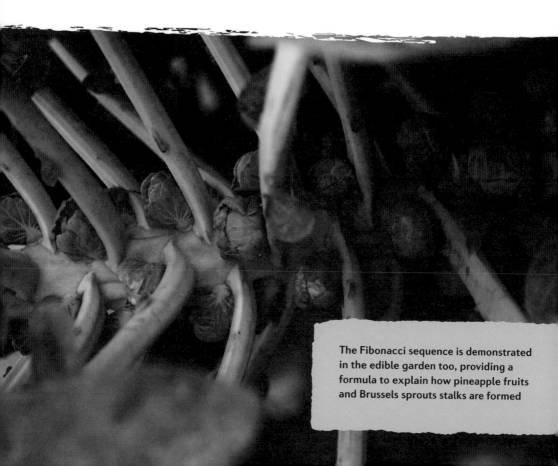

The Fibonacci sequence is demonstrated in the edible garden too, providing a formula to explain how pineapple fruits and Brussels sprouts stalks are formed

Navigate

I hope I'm never forced to rely on plants to find my way home, but if that did happen, I'd try to recall a few basic principles – things my ancestors knew, long before the advent of satnav.

Plants can only guide us because they react to stimuli, most notably heat and light from the sun to the south and prevailing winds blowing from the south-west (if you're in the UK).

Certain plants have the common name compass plant because their growth tells you which way is north. For example, the moss campion (*Silene acaulis*) produces flowers first on the south side of its cushion-like mound, to catch the most sun. The common wildflower prickly lettuce (*Lactuca serriola*) twists its leaves in response to the sun, so its margins are held upright. Unlike moss campion, it is avoiding too much sunlight, to prevent its foliage from drying out.

Trees growing on their own, say on a roadside, are typically asymmetric. The side facing the south is usually denser to gain maximum amounts of sunshine, and thereby photosynthesise. These south-side branches will tend to be growing horizontally, whereas the northern branches grow upwards. Both are searching for light.

A similar asymmetrical effect is produced when wind constantly hits an exposed tree from the south-west. If the tree is leaning towards the right, the left-hand side faces the south-west. The tree has been wind-pruned. It may also form 'guy roots' pointing towards the south-west to anchor the tree as it is pushed to the north-east.

Look out for moss and lichen on tree trunks, too. Moss, a plant that loves damp conditions, is more likely to thrive on the northern, shadier side of a tree, while lichens (which are not plants but belong to the kingdom of fungi) may prefer a sunnier spot, so take care not to confuse the two.

In autumn, leaves on the northern side of the tree tend to be shed first as they are less useful for photosynthesis. Those on the southern side will colour up quicker though, due to the effect of sunshine on the leaf chemicals responsible for autumn colour.

Plant communities are useful for indicating the presence of groundwater, proximity to the coast and soil changes too, making local knowledge of ecology invaluable.

Finally, a note of caution. Many influences other than weather, climate and soil will affect plant growth. Hills, woodland and buildings can shade the sun and deflect wind, so single plants may not give a good navigational reading. But several, showing a trend, could get you out of trouble if you get lost.

Collect

Collecting plants is a satisfying activity that can provide interesting social interaction, counter everyday stress and give satisfaction through ordering, caring and learning about the plants.

Carl Jung, the noted psychiatrist, suggested the desire to make collections is connected to our earliest origins as hunter gatherers. All these things apply to any kind of collecting activity, but special considerations are relevant to plants because they are living organisms.

There are two types of plant collecting, which have many things in common but also important differences. The first is the collection of wild plants carried out by botanists, usually for the purpose of understanding science and conserving biodiversity. The second, collecting cultivated plants, has its aim in preserving plants bred by humans for specific purposes – the prettiest, tastiest, best for medicinal use, etc.

All good-quality plant collecting involves finding, acquiring, organising, cataloguing, displaying and conserving the plants. Sometimes specimens are alive and growing in public or private gardens (living collections), but they can also be preserved as seeds, pressed herbarium specimens or as DNA.

The charity Plant Heritage, based in the UK, supports collectors of cultivated plants and awards National Collection status to the most well-kept collections. There are several categories, including taxa (botanical categories), for example all buddleias, and historical, like Queen Mary II Exoticks, which date back to the seventeenth century

Iris Xiphium. Variété

P. J. Redouté.

Langlois.

Plant collectors use illustrations
to study their specialisms and
define the categories

Iris Xiphium.

Iris Xiphium.

P. J. Redouté.

Langlois.

An iris collector has over 300 species, including this Spanish iris, to aim for

when Queen Mary imported plants from all over the world to display in Hampton Court Palace.

Of course, it's essential not to pick or dig up plants from the wild (unless you're a botanist with a permit or taking part in an official conservation project). Wild plants belong in their natural habitats, where they contribute to the ecosystem they're a part of.

The famous historical plant hunters helped themselves, but collecting wild plants is now controlled by international treaties and national laws, so modern botanists must apply for permits. This ensures plants are not over-collected – when wild populations are stripped from their native habitat.

Unfortunately, this has already happened to many species, for example the Madonna lily (*Lilium candidum*) is endangered as a result of over-collection for perfumes, medicines and gardens.

You should also take care who you buy plants from. Plant nurseries must follow government regulations to import or buy plants, but shadowy individuals on the internet selling rare plants are not so scrupulous.

Ironically, once plants have become endangered in the wild, growing them in a garden after buying from a legitimate source can be part of their preservation. The only wild population of the Wollemi pine is in a remote Australian gorge. It is critically endangered. But to conserve the species it has now been propagated and planted in botanical collections, such as Kew's arboretum, throughout the world.

One of the earliest known plant collections was that of Queen Hatshepsut of Egypt. In 1495 BCE she sent botanists to Somalia to bring back incense trees.

Photograph them

I love photographing plants because they constantly change: growing, forming leaves, flowers and fruits, or senescing, breaking and retreating back to the earth. If you record a deciduous tree each month of the year you will have twelve completely different images, even if each one is taken from the same position.

Unsurprisingly whole books, exhibitions and courses are devoted to plant photography, so this chapter will be limited to a few essentials.

Your equipment does not need to be complicated as smartphones now have cameras that can take impressive shots, especially of closeup subjects.

However, using more sophisticated equipment is fun. Detachable lenses give options like macro for flower details, wide-angle for landscapes and telephoto to capture a bird as it eats fruit at the top of a tree, for example. Resting your camera on a tripod stops it from shaking so the image is crisper.

For a good composition, check everything in the frame and not just the plant. Advice is readily available on this subject and many of the rules are easy to learn, for example three trees will often look better than four.

Try a different viewpoint – photographers at the Chelsea Flower Show often stand on stools to snap the gardens – but getting down low is great for drifts of wildflowers.

Choose your subject carefully, examining it for flaws, which are often more obvious when you look at your photo than through the lens. If you want to capture the flaws, such as dying petals, then select the best of these too!

This is easier if you're somewhere with lots of examples of the plant. So botanic and other public gardens, such as Kew, make great locations where you can pick out the perfect bloom from a flowerbed full of them. I find it hard to take good photos in a rush, so I prefer to use my own garden or a quiet location rather than flower shows with lots of people around.

Learn how to focus your camera then decide, each time, on the depth of focus whether it's a single flower or several trees. It's ok to have most of the image in high focus, but blurring the background or foreground looks dramatic and can be done even when your camera is on automatic settings.

Using a macro lens to photograph this cistus has provided sharp focus in the foreground with a pleasing blurred background

Check the subject's surroundings – a bright rose may look better set against shiny, healthy leaves than other roses.

Morning and evening give the easiest light to work with. The tones are warm and low sunlight creates highlights and shadows that help the subject look three-dimensional. Backlight, where the sun shines through flowers or leaves ahead of you, can look superb, but overcast skies are generally preferable to bright overhead sunshine for plant photographers.

I love being able to take masses of photos with my digital camera, then keeping the best ones. I'm often surprised to discover which images work out, so don't be afraid to keep shooting. It's a joy to be free from the 24 snaps available on an old-fashioned roll of film!

PLANNING A PLANT PHOTOGRAPHY TRIP

* ❋ What subject(s) are you after?
* ❋ What effect will the season have on your subject(s)?
* ❋ Research the location
* ❋ Check the weather
* ❋ Plan your journey, considering what time of day will be best for your photos
* ❋ Pack your camera, accessories, memory card and charged battery
* ❋ Wear clothes you can walk, sit and lie down in, safely and comfortably
* ❋ When you arrive, make notes or take photos to remind yourself of the plant's surroundings in case you want to photograph it again

Reduce stress and improve wellbeing

Do you feel better after sitting on a park bench at lunchtime or lying on a garden lawn?

You might think it's just the break from your routine, the solitude or perhaps the company that promotes wellbeing, but there's plenty of evidence that the plants around you are key to improving health. You don't even have to move to gain benefits to your immune system and stress levels.

We've had a hunch about this for centuries, the idea being neatly captured in the Japanese term *shinrin-yoku*, or forest bathing. In 1996 Howard Clinebell coined the word that's now familiar – 'ecotherapy' – which encapsulates the concept. Of course, animals and other organisms are part of ecosystems, but plants are essential here, whether the trees of a forest or grasses in a meadow.

Natural killer (NK) cells in our immune system protect us from cancer and viruses. NK activity has been found to increase after exposure to forests, and it's believed that 'phytoncides', plant-released substances that kill microbes (and produce the smell of pine, cypress, lavender and cedar), are responsible for this. So the benefits of simply living with plants go much further than aiding our mental health, substantially promoting physical health, too.

Cortisol is known as the stress hormone. High levels of cortisol in the body indicate exposure to stress and fear, which is great in small doses for handling difficult situations but bad for you when it's produced over long periods.

Researchers at Edinburgh University measured cortisol levels in deprived, non-working, middle-aged city dwellers. They discovered that higher levels of green space in residential neighbourhoods were associated with lower perceived stress and healthier levels of cortisol.

In another study, participants just shown pictures of nature recovered from stressors faster than those shown built environments when their heart rate, blood pressure and respiration were tested.

A pine forest is more than just a pleasant place to be, it can improve both physical and mental health

Meditate

'The ancient tradition of silent contemplation is as important to the modern mind as it was to those of our forebears – and the Quiet Garden Movement has been facilitating the practice of mindful contemplation for the past 25 years, in gardens across the world…'

These are the words of Dr Rowan Williams, a former Archbishop of Canterbury. He's talking about an organisation, the Quiet Garden Movement, which embraces a variety of traditions, churches (including the Church of England of which Williams is a bishop) and cultures.

Quiet Gardens aim to be places where people can find welcome, stillness and spiritual refreshment. There are over 300 Quiet Gardens in Europe, Africa, Australasia and North America. This linking of mindfulness, prayer, contemplation or meditation with plants and gardens has a long and deep history that covers people and cultures from all around the globe.

Flowers are often used to adorn temples in many Buddhist and Hindu traditions. Within the Thai Buddhist tradition temples are set in gardens, which often include frangipani for scent and lantana to attract butterflies and other insects. In China, nandina was planted on temple grounds as it symbolises good fortune, and this association has led to its common name, heavenly bamboo.

Within the Tibetan Buddhist tradition flowers are occasionally used as part of Meditations on Impermanence, contemplating that they are beautiful but that they will inevitably wither and die.

The flower of the lotus (*Nelumbo nucifera*) is central to the Sanskrit mantra, *Om mani padme hum*. The lotus signifies purity, due to its ability to emerge unstained from the mud in which the plant grows.

Thich Nhat Hanh, a Vietnamese Buddhist master, taught a walking meditation called 'A Flower Blooms beneath Each Step' where practitioners were instructed to visualise a lotus blooming under each step of the walk the moment their foot touched the ground.

In Tibetan Buddhism and within the Sutras themselves there are references to a number of different lotus plants. These include *Nelumbo nucifera*, often seen growing in ponds around Buddhist temples where the climate will allow, but also the utpala flower, which is blue in colour and generally recognised as *Nymphaea caerulea*.

Aromatic plants are burned to provide scented smoke by the indigenous peoples of America, using white sage (*Salvia apiana*) for example. Further east, frankincense, the resin of various *Boswellia* species, has been traded for over 6,000 years for medicinal and ceremonial use and is still burned as incense in Christian and other religious ceremonies.

The Japanese tradition of arranging flowers, stems and leaves has a contemplative element not usually found in European floristry. In *ikebana* the rules of arrangement are designed to preserve the life of the flowers, with the surface of the water exposed in a wide-necked vase. Some practitioners feel silence is needed while carrying out *ikebana*.

Supply the kitchen

Make wine

Making high-quality wine from grapes is a fine art, perfected by experts over millennia. But the fact that drinking good wine was clearly a standard element of an ordinary wedding in biblical times reminds us that the production of drinkable wine is not the mysterious art of a genius but a basic technique. Certainly, a Mediterranean climate helps, but sparkling white wines from England have won international awards and I can testify that they're delicious.

I wasn't convinced an ordinary domestic garden in England could produce pleasant wines until I visited a keen home gardener (Bob) in Norfolk on a sunny summer day to learn the basics.

Wine comes from varieties of *Vitis vinifera*, a vigorous climbing plant that plunges its roots deep into the ground, enabling it to withstand hot, dry, Mediterranean summers. Choose one or two varieties recommended for your region.

Bob suggests growing one in a polytunnel and one outside, to balance the flavours. Prune and train your vines to produce enough, but not too many bunches of grapes, each one exposed to plenty of sun. This means wielding the secateurs several times a year!

Harvest the bunches at the end of summer, remove any rotten grapes and press the clean ones, then ferment the juice in demijohns, with a starter yeast. During fermentation, the sugars in the juice are converted to alcohol, which remains in the liquid, and carbon dioxide, which is released.

Added sugar, once fermentation has slowed, increases strength, then after a month, the liquid is decanted and stored in bottles until it's ready for drinking after about a year. This method makes white wine but red wine is only a little more complicated, with the skins also squeezed and fermented to add colour and tannin flavours.

Plants are not the only sources of alcoholic drinks – fermented milk or honey drinks are popular around the world – but the animals from which those ingredients come must feed on plants while beer, whisky, vodka, gin, rum and tequila etc. are all plant-based.

Making grape juice is as simple as squeezing the fruits, but wine production needs only a few more steps

Seal/cork bottles

One of the most intriguing things to find in the Iberian countryside is a group of recently cut cork oaks, the grey gnarled bark of their limbs and trunk ending suddenly in a neat cut, below which is what looks like bright red, bare wood.

On the ground, aromatic herbs and scented flowers like lavender and cistus abound, because this isn't intensive farming but an ancient, bucolic ecosystem where these remarkable trees are allowed to flourish along with the plants and animals that thrive beneath and around them.

Of all the things you can do with plants in this book, this section probably contains the easiest things, because simply choosing a bottle of wine with a real cork provides support for a fascinating crop and a threatened ecosystem.

The cork oak, or *Quercus suber*, gives its name to suberin, an elastic but impermeable compound contained in large quantities in its corky bark. Suberin is key for these trees to survive forest fires. Protected by their insulating bark, cork oaks remain standing through a conflagration, ready to regrow their burnt branches.

Surprisingly, and very unusually among trees, cork oaks can also survive the removal of the outer layer of their bark by cork farmers, several times during the trees' lives. This process is skilled, involving axes rather than machinery, with the operators using considerable force to cut the 'planks' of bark, while not damaging the underlying phellogen (the layer of live cells from which new bark grows).

Cane toppers protect gardeners from eye injuries and are just one of hundreds of functions for the bark of cork oaks

The process begins when the trees are 25 years old, and continues, every 9–12 years, until the trees are around 250 years old. As the bark regrows, it turns from red to black. Sometimes you can see numbers painted on the trunks, indicating when they'll be ready for cutting again.

Farming cork oaks is an extremely sustainable activity, with the trees remaining in place for centuries while forage crops may be grown beneath them for livestock. Often the trees are not accessible to vehicles so the cork planks must be carried away by hand.

Cork oak forests are diverse ecosystems, supporting endangered species of plants and animals, such as Iberian lynx in the 'dehesas' of

Spain, and Barbary macaques in the Atlas Mountains. After use, cork can be recycled or composted, eventually breaking down in a way that synthetic substitutes cannot. One study has shown that plastic stoppers release 10 times more CO_2 than cork ones and aluminium stoppers/caps 26 times more.

Cork stoppers have their critics, being blamed for cork taint, an unpleasant flavour caused by a chemical reaction that affects a small percentage of wine bottles. However synthetic options don't allow wine to age into a vintage product.

It's the combination of impermeability combined with a tiny amount of breathability (called micro-oxygenation) that makes natural cork perfect. Champagne producers use the second cutting of the bark, when it has reached exactly the right stage for their premium product.

It's not surprising that this miracle material is also used in the manufacture of other niche products – a coating of cork compound protects spacecraft from the 1,000°C temperatures they are subjected to on lift-off and return to Earth. In the car industry cork provides soundproofing. It also has specialist uses in sports such as cricket and fishing and more everyday applications in flooring and other building work.

Despite being so beautiful and useful, cork oak forests are under threat from poor land management, climate change and pathogens, so it's vitally important that they are valued and protected. One of the best ways of achieving this is through the utilisation of cork in the wine industry. By choosing a bottle with a natural cork, consumers can vote for sustainable farming.

Numbers painted on cork oak
trunks indicate when they'll be
ready for the next bark harvest

Make a hot drink

Making a tasty hot drink is one of the easiest things you can do with plants. What's more, the ingredients are easily grown in your garden if, like me, you're a fan of a tisane.

Tisanes are drinks made with nearly boiling water, by infusion or decoction of plant material. The simplest form is a handful of mint leaves placed in a mug onto which water from a kettle that has recently boiled is poured. This is so easy to do that you might wonder why anyone with a suitable garden ever buys mint teabags.

There are a few (minor) drawbacks to growing and making your own mint tea. Mint of many kinds can be used but *Mentha spicata*, or spearmint, is particularly tasty. This herb is vigorous to the point of being invasive in moist soil. It produces an abundance of underground and overground stolons (stems or runners that take root) so easily takes over a herb bed.

An easy solution is to grow it in a pot, either free standing, or with the pot buried in a bed. The stolons will still be produced, but can't progress below the soil surface, and above the surface can easily be pulled up and cut back to the container edge. In my garden, mint dies back to the ground in winter and resprouts in spring, but I can keep it producing leaves all year round by growing a pot in my unheated greenhouse.

Some people don't like large quantities of leaves floating in their drink, but that problem is easily resolved with a teapot, and after drinking, the dregs make perfect material for a compost heap.

I've never tried being self-sufficient in ingredients for tisanes, but enough plants for a year's supply would take up a lot of space. Don't let all of that put you off though – there is no mint tea quite as flavoursome as one brewed from a big bunch of fresh leaves.

This method applies to parts – leaves, fruits, roots, seeds etc. – of many other plants, a non-definitive list of which you can see on page 132. These have all been used to make refreshing drinks for centuries. Some are reputed to give medicinal benefits, but none contains caffeine. You can make a summer drink by allowing the infusion to go cold before serving with ice, or sweeten your tisane with sugar or honey.

Rosehips, rich in vitamin C, lycopene and other carotenoids are popular ingredients in 'teas' and syrups

Generally, making caffeinated drinks involves techniques that an amateur would struggle with. As a rule, the plants containing caffeine grow in tropical regions – *Coffea, Theobroma, Cola*. Types of tropical holly and *Paullinia* contain it, as well as *Camellia sinensis* – the source of Indian and Chinese tea. The latter grows happily in the UK, with one farm being situated as far north as Perthshire in Scotland.

Making a cup of tea from C. *sinensis* is more complicated than making a tisane, simply because there are several steps – you pick and steam the leaves, then rinse under cold water. Next you roll and dry them in your oven and finally place them in a teapot to steep in just off-boiling water. Worth a try, but one bush won't supply most tea drinkers' needs for long!

SOURCES OF TEAS AND TISANES

❀ Apple	❀ Lemon	❀ Pelargonium
❀ Bergamot	❀ Lemon balm	❀ Pine
❀ Chamomile	❀ Lemon grass	❀ Rose blossom
❀ Chrysanthemum	❀ Lemon verbena	❀ Rosehips
❀ Echinacea	❀ Liquorice root	❀ Rosemary
❀ Elderberry	❀ Mallow blossom	❀ Tulsi basil
❀ Elderflower	❀ Marigold petals	❀ Turmeric
❀ Fennel	❀ Mint	❀ Valerian
❀ Ginger	❀ Nettle	❀ Vetiver
❀ Hawthorn	❀ Olive leaf	
❀ Lavender	❀ Orange zest	

Smoke food

Smoking food, to flavour, cook or preserve it, dates back to the Paleolithic era. The smoke acts as an antimicrobial and antioxidant, but by itself is insufficient for preserving meat and fish as it doesn't penetrate far enough into them. To preserve the food as well, it must be heated, so a hot smoking method, where both heat and smoke are generated, will cook the food to the point where it can be kept for longer than raw food.

Around the world various types of wood are used, with alder (*Alnus glutinosa*) being a traditional European choice, along with oak or beech. While smoking by burning dung or peat is sometimes done, tree wood from hardwoods (trees that are not conifers) is the usual source of smoke. The resin from softwoods produces too much soot.

Burnt lignin yields pungent compounds like guaiacol, phenol and syringol, while sweeter scents may come from vanillin or isoeugenol. Proteins in wood can produce roasted flavours. The sugar molecules in cellulose and hemicellulose caramelise when burnt to give colour and sweet, flowery or fruity aromas. Phenolic compounds slow bacterial growth and rancidification of animal fats. Other active antimicrobials found in wood smoke are formaldehyde and acetic acid.

The simplest way to smoke food is called trench smoking. A narrow trench is cut down a slope, pointing into the prevailing wind. The middle part of the trench is covered over. At the upper end a chimney is formed, with a rack in it on which the food to be smoked can be placed. At the lower end of the trench, the smoky fire is lit and sustained day and night until the food is cured.

Amateurs are not advised to try this as badly smoked food will not be adequately preserved. There are also some health concerns around carcinogens in smoked food, particularly where the history of the burning wood is unknown.

Trees in the Fagaceae family, such as this beech, are popular for food smoking

Make a pudding

Plant-based ingredients are vital for puddings. The sweet element is provided by sugar, plant-based syrups or occasionally honey (made by bees but impossible without flowers for them to forage upon). Any stodgy element will come from grains, such as wheat, nuts or root crops. Your recipe may also be invigorated with a strong flavouring such as lemon, vanilla, cinnamon or ginger – all, once again, derived from plants.

While it's easy to buy your sweetener, starch and spices from the supermarket, it's much more satisfying to grow a pudding yourself! You'll need to buy one or two of the ingredients for the dishes I'm suggesting, but the bulk of them can come from your plot.

Favourite home-grown ingredients include carrots, beetroots and courgettes, which provide both sweetness and moisture. Mature carrots and beetroot, for example, contain over 7 grams of sugar per 100 grams.

The scent of alder wood (left) is claimed to provide a sweet and subtle flavour to smoked food. The vanilla plant (right) is a tropical epiphytic orchid grown for its 'vanilla beans'

For **carrot flan**, mix 450 g (1 lb) grated carrots, 4 drops almond essence, 4 tablespoons plum jam and 4 tablespoons of water. Simmer to a thick pulp for 20 minutes, then spoon into a cooked pastry case and bake for 5 minutes at 180°C (350°F). Carrot flan was a ration-book wartime favourite, along with beetroot pudding – an impressively pink dish!

For **beetroot pudding**, mix 170 g (6 oz) plain flour, ½ teaspoon baking powder, 30 g (1 oz) butter, 30 g (1 oz) sugar and 120 g (4 oz) finely grated beetroot. Bind with a little milk into a cake-mix consistency. Add a few drops of banana essence and bake at 180°C (350°F) for 35 minutes. Serve both of these recipes with ice cream or custard.

To make **courgette muffins**, grate 50 g (2 oz) courgette and an apple. Add the juice of half an orange, one egg, 75 g (3 oz) butter, 300 g (10 oz) self-raising flour, ½ teaspoon baking powder, ½ teaspoon cinnamon and 100 g (4 oz) caster sugar. Spoon the mixture into a greased muffin tin and cook for 25 minutes at 190°C (375°F). Ice your muffins for an extra treat! If you have watery courgettes, squeeze them in a clean cloth and discard the moisture.

Try making **a fool** from allotment fruits by simmering 400 g (14 oz) with 125 g (4 ½ oz) caster sugar and 75 ml (3 fl oz) water. Strain off the bulk and mash it, then reduce the remaining syrup by a third. Whisk

CARROTS

BEETROOT

300 ml (½ pt) double cream with 500 g (18 oz) yoghurt plus vanilla extract, mix in the fruit bulk and serve with the syrup.

Most allotment fruits (rhubarb, gooseberries, redcurrants, blackcurrants, apples and pears) work perfectly in the fool recipe or in pies and crumbles.

To get the best out of allotment ingredients, learn how to preserve or store them for less abundant times.

Keep apples and pears in a cool place on racks, and freeze rhubarb, gooseberries, redcurrants and blackcurrants. Use up courgettes in cooking but you can freeze the muffins you make. To get the benefits from carrots and beetroots over winter you can leave them in the ground until you need them.

These dishes all contain animal-based ingredients, but you can substitute these with coconut or oat milks if you want a purely vegan (so purely plant-based) pudding.

For vegan chocolate, coconut and caramel pots, blend coconut milk, dates and vanilla bean paste, then simmer for 10 minutes. Divide into glasses and chill. Next put vegan chocolate in a mixing bowl, heat oat cream and maple syrup, and pour over the chocolate to melt it. Whip more oat cream and blend it into the melted chocolate, then spoon it over the coconut caramel – delicious!

COURGETTES

GOOSEBERRIES

Wrap food

What if there was a wrapping material that was renewable, biodegradable and needed no industrial processing? Step forward the leaf. Wrapping food in a foliage package for cooking protects it from direct flames, holds in heat and allows it to steep in its juices.

Sometimes culinary leaf wrappings are composted or given to animals to eat afterwards, but they may form a delicious part of the dish, as with the grape-vine leaf wrappings of the Greek dish dolmades. Banana leaves are particularly versatile, they're water resistant and provide a subtle flavour.

Simply heat until they're pliable in an oven or boiling water, and to tie your package use strips of leaf. Unprocessed, leaves of bananas (such as *Musa* x *paradisiaca*) don't last long but with a new technology, called cellular enhancement, they are said to be useful for up to around three years.

Have you ever seen a droplet of water rolling on the surface of a lotus (*Nelumbo nucifera*) leaf? 'Ultra-hydrophobicity' prevents water from settling or soaking into the leaf – it simply forms round droplets that shimmer on the surface like glass baubles. Studies have shown that leaf powder from lotus plants increases refrigeration time limits, so these leaves appear to have other useful applications in the kitchen. These qualities make the leaves ideal for wrapping chicken.

Another popular tropical plant for leaf cooking is the screw pine or pandan (*Pandanus amaryllifolius*). After blanching, the leaves are rolled and woven into cylinders to contain, flavour and colour rice.

Not really a wrapping, but worth a mention as watching them being made is engrossing, are the sturdy plates used in Indian festivals. These plates, called patravali or khali, are fashioned from large leaves of the sal tree (*Shorea robusta*), stitched together and then moulded onto biodegradable cardboard. Banyan (strangler figs) and other local large-leaved species may also be used.

And while there's a long tradition of cooking with big leaves from tropical regions, there's no reason why British cooks can't also get creative with home-grown Swiss chard, grape-vine, cabbage, horseradish, common fig or lettuce leaves.

Chard is related to beetroot and spinach, easy to grow and produces huge leaves suitable for making savoury food parcels

Inspire creativity

Fashion furniture

While there are numerous different types of furniture, there are even more types of plants from which it can be made – think hammocks made from any of the rope plants (see page 145), bamboo screens, rattan and wicker chairs, or wooden tables.

It's thought that the earliest furniture item was the kind people sat on – probably the simplest type of seat – a stool. Even stools can be made from many plants – woods from trees and shrubs or hulms from bamboos, with solid seats or those woven from any of the numerous fibre plants. You might just sit on a block of wood, or a pouf crocheted with cotton or hemp yarn.

I naturally think of wood if furniture is mentioned, a material that is as complex as the number of tree and shrub species in the world, each one having its own characteristics due to the way it grows.

The living part of a tree trunk is on its outside, just beneath the bark. Here, in the newest wood (sapwood), are the vertical vessels that transport food and water to its canopy. The centre consists of old, redundant vessels (heartwood), no longer needed for transport to the point that they may decay altogether, leaving a living tree completely hollow.

In the spring, the vessels contain large cells, to carry lots of sap. In autumn, thinner cells carry moisture upwards. These differences make patterns in wood you can see if you cut down a tree.

How to make a stool

1 Prepare three legs of the same length by shaving them down from a slim branch

2 Allow the legs to dry so that they shrink down to their narrowest width

3 Make tenons (pegs) on one end of each leg – these will be inserted into the seat

4 Saw a small (bottom-sized) plank from a tree trunk or large branch

5 Before the plank has dried, make three equidistant mortise holes in it the right size to fit the tenons

6 Bang the tenons into the holes

7 As the seat dries it will gradually contract, and hold the legs tightly

Green woodwork relies on the unseasoned wood shrinking to form tight joints

Woods are tested for desirable characteristics to use for different purposes, for example, their moduli of elasticity and rupture, Janka hardness (how easy it is to push a steel ball into its surface) and crushing strength. The latter is particularly useful to know for chair, table and stool legs. It is the wood's strength when weight is applied to its ends.

Wooden furniture is made from a renewable material that is a carbon sink and can be composted at the end of its life, but forests must be managed sustainably (see page 16) with new plantings replacing felled trees and careful choice of the most appropriate trees to plant. Some wood preservation techniques can harm the environment while storage and transport must be managed to avoid waste and the spread of pests and disease. Also, many desirable timbers, such as mahogany (*Swietenia*) come from trees that are now endangered due to unsustainable logging.

Make rope
and cordage

Although modern rope is usually made with synthetic materials, there is still a place for natural rope made from plants. It's aesthetically pleasing, biodegradable at the end of its life and less slippery than synthetic rope, which (like other synthetic materials) can also be damaged by UV light.

As recently as the 1970s in the UK, tree-climbing ropes for arborists were made from hemp, which is still used for climbing equipment in gyms.

Along with hemp from the plant *Cannabis sativa*, 'Manila' hemp or abaca from the banana relative *Musa textilis* is a traditional material for rope-making.

Coir, sisal, linen (flax), cotton, jute and straw continue to be popular plant-derived materials, providing fibres that are both strong and flexible. These are processed in a similar way to paper fibres (see page 159). Ropes are made of twisted strands, which ensures tension is distributed evenly. The strands themselves are formed of twisted yarns, and these are spun from plant fibres. At each stage of the process, the materials are twisted in opposite directions, leading to a stable and unified rope.

The basic techniques of rope-making can be adapted to make cordage for bushcraft, basket making and garden projects. Fibrous plants like date palms, grass and flax are traditional materials but iris, crocosmia, daffodil, rush, stinging nettles, phormiums (New Zealand flax) and yuccas work well too.

Or try raffia (from the tropical palm *Raphia*) – it's a great material to practise with, especially if you have two colours.

How to make cordage

1 Gather a few strands of each of orange and neutral fibres

2 Knot them together at one end

3 Pull the orange strands to the right and twist them twice to the right

4 Cross them over the neutral stands

5 Repeat the process with the neutral strands

6 Keep twisting to the right and crossing to the left, adding new strands to make the cord longer

7 Knot the end to keep your work in position

Natural cordage is easy to make from raffia or other plant material and is useful, strong and attractive

For best results, harvesting of rose flowers for oil should be carried out at dawn and the material distilled the same day

Make perfumes

If you'll excuse the puns, it's hard to distil the art of making perfume down to its essence in a few words! However, put simply, plants provide vitally important ingredients in the perfume industry and have done so for thousands of years. The perfumes are made using every part of plants, from roots to seeds.

Sophisticated perfumes include elements that last a long time: base notes (for example tobacco), top notes, giving an immediate impression (such as lavender) and middle notes, in between (like sandalwood). Another wood commonly used in perfumes is agarwood (a product of mould-infected aquilaria trees). Bark from cinnamon and cascarilla is popular, along with the leaves of patchouli, rosemary and even tomatoes.

The essential oils and aroma compounds provided by plants for perfumes are often also their way of attracting pollinators or protecting themselves from herbivores.

Scents are usually extracted by maceration, submerging woody or fibrous plant parts in a solvent for anything up to a few months, or distillation whereby petals, for example, are heated and their fragrant compounds collected after condensing the distilled vapour. Citrus peels are simply compressed until their essential oils run off. These aromatics are then blended with a solvent, usually ethanol, which itself derives from plant sugars.

How to make rose-petal perfume

You can buy an essential oil still to make a distillation that will stay fresh for several months. Alternatively try this simple method:

1 Drape a cheesecloth over a bowl

2 Place rose petals on top of the cheesecloth

3 Cover the petals with a little water, cover with a lid and leave to soak overnight

4 Lift the cloth containing petals and water and squeeze over a saucepan

5 Simmer liquid down to around a teaspoon. This can be stored in the fridge and used for about a month

Attar of roses is the essential oil extracted from the petals of various types of rose

SOME IMPORTANT PERFUME PLANTS

GENUS	SPECIES	COMMON NAME
✳ *Abelmoschus*	*moschatus*	muskmallow
✳ *Acacia*	*dealbata* and *farnesiana*	mimosa and cassie
✳ *Aquilaria*	many species	agarwood
✳ *Cananga*	*odorata*	ylang-ylang
✳ *Chrysopogon*	*zizanioides*	vetiver
✳ *Cinnamomum*	*camphora*	camphor oil
✳ *Cistus*	*ladanifer*	labdanum
✳ *Citrus*	*aurantium*	petitgrain and neroli
✳ *Citrus*	*bergamia*	bergamot
✳ *Croton*	*eluteria*	cascarilla
✳ *Dipteryx*	*odorata*	tonka bean
✳ *Jasminum*	*grandiflorum*	grasse jasmine
✳ *Jasminum*	*sambac*	Arabian jasmine
✳ *Juniperus*	*communis*	common juniper
✳ *Lavandula*	many species	lavender
✳ *Narcissus*	*poeticus* and *tazetta*	aobacco
✳ *Pinus*	many species	pine
✳ *Pogostemon*	*cablin*	patchouli
✳ *Polianthes*	*tuberosa*	tuberose
✳ *Rosa*	many species	rose
✳ *Salvia*	*rosmarinus*	rosemary
✳ *Santalum*	many species	sandalwood
✳ *Simmondsia*	*chinensis*	jojoba oil
✳ *Solanum*	*lycopersicon*	tomato leaf
✳ *Syzygium*	*aromaticum*	clove buds
✳ *Vanilla*	*planifolia*	vanilla orchid
✳ *Viola*	*odorata*	violet

Preserve them

From bunches of scented lavender through natural confetti to crystallised primroses, preserved flowers can be useful and attractive as well as making your garden bounty last longer.

There are numerous preservation methods, each one suiting a different group of plants. Most simple is air drying by hanging flowers and foliage upside-down until they have lost all moisture. Cut the flowers just before they're fully open or cut seed heads and foliage before they can be damaged by weather.

Choose a drying spot that is warm, dry and dark. For flowers with weak stems attach wire to the flowerhead before hanging. Remove any foliage from flower stems.

Ideal flowers to dry this way include limonium, lavender, xerochrysum, heather, peonies and roses, and for seed heads, *Scabiosa stellata*, ornamental grasses, physalis, opium poppies and nigella. Flowers should be dry and brittle before you take them down.

For confetti leave petals on cardboard and keep in a warm, dry room. Some people speed up the process using a slow oven.

For a more supple and longer-lasting effect, steep foliage and certain flowers in glycerine (also called glycerol or glycerin). This works well for foliage and flowers like hydrangea, bells of Ireland and gypsophila.

Mix two parts warm water with one part glycerine and stand the flower stem or submerge the leaf in it. After two to three weeks the

plant vessels will be stiff and self-supporting, having taken up the gycerine to replace their water content.

Hydrangeas can also be dried by standing them in pure water. They do not dry well when hung upside-down. Cut them on a dry day and strip the leaves from the stems. Place their vase somewhere cool and dry, then simply wait for the water to evaporate.

To make crystallised flowers for cake decorations, choose edible flowers, such as primroses, violets and rose petals, that you are sure have never been treated with pesticides. Dip them in beaten egg white, then sugar and leave to dry on baking paper for a couple of hours. After that they can be stored in an air-tight container for a month or so.

Herbariums use several methods, including pressing, to preserve plant material for posterity so it can be examined by botanists. Some of the herbarium specimens at Kew are 170 years old! Pressed herbarium specimens can be quite bulky and are usually brown. The plants are not arranged to show their beauty but so that people studying them can count their plant parts and measure them.

Using an ordinary flower press, however, you can create beautiful pictures with much of the colour retained. Best results are from fresh, dry flowers, pressed straight after picking and carefully arranged so, if possible, the petals are not folded.

Keeping to one thickness of material in the press, for example by separating petals from their stems. The press squeezes the flowers between sheets of blotting paper by turning screws (or you can use a heavy book). You should have some beautiful results within around three weeks.

Weave a basket

Basketry techniques are used all over the world, adapted to employ the different plants available and objects needed. Plant materials must be pliable and strong and preferably long, so little joining is needed, as this weakens the article and slows the process.

Materials that are fine, short or weak, like grasses, can be bound or plaited together before weaving. Tree trunks, on the other hand, although long and strong, need splitting into pieces that are fine enough to be pliable.

In the UK, hedgerows are good sources of colourful materials (or 'stuff') that are easy to work with (see page 158). Most need gathering between November and March, when the sap is not flowing, then leaving in a sheltered outdoor spot for around three weeks to be partially dried or 'clung'.

The stuff should then remain useful until April, by which time it will be too dry. Look for rods that are long, straight (without side shoots), no thicker than a pencil, pliable enough to bend round your fist, only one year old and lacking much pith.

If you come across fruits like these sloes in the wild, collect them quickly in a speedy basket fashioned from brambles

How to make a forager's basket

This is an easy beginner's basket, which can be made on site and lined with leaves before you forage. If you use lengths of this year's bramble it doesn't need to be clung but will only remain supple for a couple of days. You will need gardening gloves and secateurs.

1 Cut ten 'weaver sticks' – 1.5 m (5 ft) lengths of fine bramble stems

2 Cut five 'base sticks' – 30 cm (12 in) lengths of slightly thicker bramble stems

3 Remove any thorns by running your gloved hands down the lengths

4 Weave the five base sticks into a flattish, evenly spaced under-and-over grid, like a noughts-and-crosses grid with an extra side

5 Line up the first weaver stick with the two parallel base sticks, then weave its long end over and under the three parallel base sticks

6 Continue circling round the base, weaving under and over the ends of the base sticks in the grid and when you have come full circle, keep going round

7 On around the fourth circuit, pull up the arms of the base sticks to form the sides. (This is called 'randing'.) Keep circling the weaver stick inside and outside the uprights

8 You can add extra length to an upright by inserting a new stick next to the short upright

9 To join a new weaver stick, just cross it over the previous weaver stick next to one of the uprights. This looks neat if done on the basket's inside

10 Push down on the rows of weaving occasionally, to close any gaps

11 Finish weaving by pushing the end of the last weaver stick under the weave in the row below

12 Trim off the tops of the stakes

HEDGEROW PLANTS FOR UK BASKET MAKING

- Bramble
- Buckthorn
- Bullace
- Clematis
- Cornus
- Dog rose
- Hazel
- Honeysuckle
- Hornbeam
- Ivy roots
- Jasmine
- Lilac
- Lime
- Periwinkle
- Poplar
- Privet
- Snowberry
- Spindle
- Spiraea
- Virginia creeper
- Willow

Basket weaving has many useful applications in the garden, such as tunnels, fences or bed edges

Make paper

Making your own paper is easier than you might imagine – people have been doing it for thousands of years. You don't need to cut down a tree as some of the best plants to use are not woody plants, but the taller, more substantial, monocotyledons, like bananas, corn, bamboo, bulrush, seagrass, iris, lily or sugarcane.

If you'd like to try with one of your garden plants, aim for one that stands on its own at a metre or more high.

There are several steps to the process, so it's worth gathering your equipment before you start. Boil your plant material with washing soda. This will draw out impurities, leaving behind the cellulose fibre you need.

Next, beat it. A hand blender is the lazy tool, while a mallet gives the good, long fibre that will make a smooth, strong paper. (In elephant sanctuaries this part of the process is done in an elephant's gut and the resulting 'elephant-poo paper' sold for conservation funds.)

Place the sludge you've made on a mould and deckle. This is a sieve-like screen, through which the liquid gradually drains, leaving behind randomly interwoven fibres, called furnish.

Transfer the furnish to an absorbent surface – try a flower press – then allow to dry. Paper made like this is characterful and rustic, whereas commercial paper production generally involves harsh chemicals, lots of water and clear-cut forestry.

Look out for plants with species names like *papyrifera* (e.g. *Broussonetia*, the paper mulberry) or *papyracea* (e.g. the daphne from which lokta paper is made). These are the sources of traditional, refined papers, but other plants have also been used as surfaces to write on through the years, including birch bark by guerrilla fighters in World War Two.

Papryrus is made from the inner pith of the sedge, *Cyperus papyrus*. Thin strips are placed side by side with overlapping edges and another layer placed on top at a right angle. These are hammered together, dried and polished.

The popular houseplant, *Cyperus alternifolius* is related to *C. papyrus*, the source of paper in biblical times

Weave a floor mat

Floor mats offer comfort by softening hard and cold floor materials as well as enhancing the look of a room, or simply delineating areas.

The techniques for making them are numerous, but weaving, knotting, crochet and plaiting techniques are the most popular ones for plant-based mats. Floor mats made from plants can be composted when they become worn out or dirty. If made from ecologically farmed materials, they will have little negative impact on the environment, while the crop will capture carbon as it grows and store it afterwards.

The traditional Japanese mats called tatami are woven from cotton and rush with a core of rice straw. The size of a floor may be calculated by the number of these mats needed to cover it. To keep the mats clean, visitors remove their shoes before stepping on them.

Doormats in the UK, on the other hand, are designed to remove dirt from shoes and are usually made from coir (coconut palm fibres), cane, hemp, grasses or rushes. Other palms, such as the palmyra palm, sedges, flax and even the inner bark of lime trees also provide popular matting materials.

Jute has increased in popularity for mat making as its production is largely environmentally friendly. It comes from the genus *Corchorus*, in the same family as mallows and abutilon.

The long bast fibres (from the food-carrying phloem vessels in the stem) are harvested after around four months of summer growing

How to make a jute mat

1 Buy rolls of 2 mm organic jute

2 Build a loom by nailing two rows of finishing nails into the top and bottom of a board around 1 cm (1/2 in) apart

3 Run a warp thread across the whole board from top to bottom to top etc. and fix both ends tightly

4 Thread a tapestry needle or stick shuttle with a length of yarn and thread it under and over the warp thread, taking care to secure the end

5 Weave back in the other direction, taking the weft thread over the warp threads you previously went under and under those you went over before

6 Continue weaving and push down the weft threads so they are tight with a comb or fork

7 When you've woven to the top of the board, pull the warp rows off the nails and tie adjacent pairs together

8 Repeat at the base

in humid, tropical regions. If grown in suitable climates, jute can be farmed organically as part of a crop rotation, so much of the world's supply is organic.

Once harvested, like flax (see page 72) jute must be retted to extract the fibres, and this process can be done biologically (organically) or with synthetic chemicals.

Paint/draw them

Drawing flowers is a hobby enjoyed by 3 to 90-year-olds and beyond, with results varying from doodling to intricate portraits useful to botanists. Artists creating the latter are professionals, trained not only to wield the brush or pen with skill but also to understand plant anatomy at a high level. Their illustrations are so precise and intricate that they're better than photographs as a record and reference of a plant's characteristics.

Consequently, botanic artists are employed by botanic gardens like Kew to assist with taxonomic classification of plants. Their work helps herbariums compare species and even describe completely new ones. An artist's plate (illustration), published alongside a written description, is part of the definition of a plant, a tool of science and communication.

Plant material brought to botanic gardens from expeditions is preserved for the journey, often as a dry, brown, pressed herbarium specimen (see page 153). Botanical artists bring it back to life, mixing colours and measuring the plant parts to accurately record its appearance.

In the plates they create they will have angled the parts of the plant so the unique elements of its structure can be seen. For example, a flower squashed sideways in a press may be turned head-on with its stigma, stamens and petals positioned ready to be counted. The artist uses a microscope to explore minute details, such as the position of hairs.

While the work of botanical artists is inspiring and admirable, it's just one of the infinite ways to make art from plants. Van Gogh was not limited by measuring tools and microscopes when he painted his sunflowers, but they continue to inspire generations over 130 years later!

WHAT TO USE

Ways to make floral illustrations – the results will vary depending on your materials. Here are four colourful ones suitable for use on paper with ideas to get the best out of them:

COLOURED PENCILS
You can sharpen a pencil to a precise point so these are good for accurate representations of a plant's details. Some are water soluble, meaning you can intensify the colour on the page or create a wash effect like watercolour.

ACRYLIC PAINTS
Like oils but easier to use because they dry faster. Great for paintings of gardens or vase arrangements when applied impressionistically to communicate a feeling. Can be layered.

WATERCOLOUR PAINTS
A favourite of botanic artists, good for both precise details, with tiny brushes, and bold washes over large areas. Usually combined with pencil outlines. Can also be layered.

FELT-TIP PENS
Perfect for design work, making patterns or colouring in. Precise, bold and quick, with no need to mix colours.

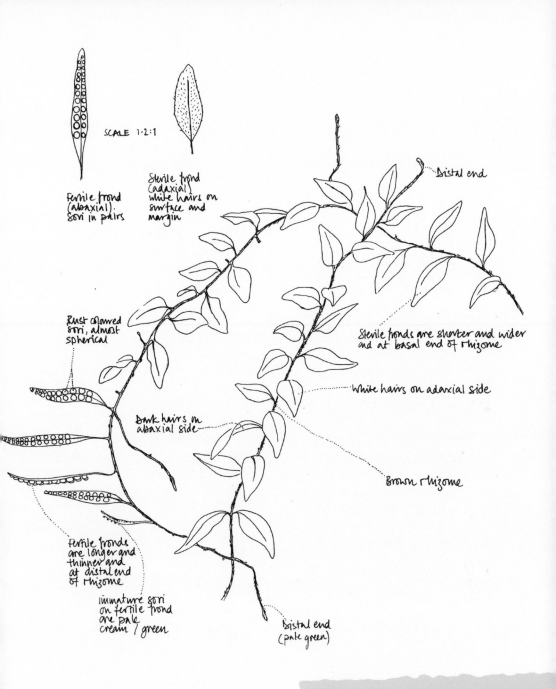

SCALE 1·2:1

Fertile frond
(abaxial).
sori in pairs

Sterile frond
(adaxial)
white hairs on
surface and
margin

Distal end

Rust coloured
sori, almost
spherical

Sterile fronds are shorter and wider
and at basal end of rhizome

White hairs on adaxial side

Dark hairs on
abaxial side

Brown rhizome

Fertile fronds
are longer and
thinner and
at distal end
of rhizome

immature sori
on fertile frond
are pale
cream / green

Distal end
(pale green)

A diploma project brought out the artist
in me: this is my drawing of an unusual
fern called *Microgramma reptans*

Further reading

Beeton, I. (1960). *Mrs Beeton's Cookery and Household Management.* WARD LOCK AND CO., LONDON.

Berners-Lee, M. (2020). *How Bad are Bananas? The Carbon Footprint of Everything.* PROFILE BOOKS, LONDON.

Blackmore, S. (2018). *How Plants Work: Form, Diversity, Survival.* IVY PRESS, BRIGHTON.

Gibbs, N. (2012). *The Real Wood Bible: The Complete Illustrated Guide to Choosing and Using 100 Decorative Woods.* FIREFLY BOOKS, RICHMOND HILL.

Heap, T. (2021). *39 Ways to Save the Planet: Real World Solutions to Climate Change – and the People Making Them Happen.* WITNESS BOOKS, LONDON.

Hecht, A. (2001). *The Art of the Loom Weaving: Spinning and Dyeing Across the World.* BRITISH MUSEUM PRESS, LONDON.

Johnson, K. (1991). *Basketmaking.* BATSFORD, LONDON.

Mabberley, D. (2017). *Mabberley's Plant-Book: A Portable Dictionary of Plants, their Classification and Uses.* CAMBRIDGE UNIVERSITY PRESS

Miles, A. (2021). *The Trees that Made Britain.* PENGUIN BOOKS, LONDON.

Polunin, M. and Robbins, C. (1992). *The Natural Pharmacy: An encyclopaedic Illustrated Guide to Medicines from Nature.* DORLING KINDERSLEY, LONDON.

Robinson, R. (1973). *The Penguin Book of Sewing.* PENGUIN, LONDON.

Van Wyk, B. (2019). *Food Plants of the World: An Illustrated Guide.* TIMBER PRESS, PORTLAND.

WEBSITES

amnh.org/research
bbc.co.uk/news/science-environment
bbcgoodfood.com
climatekids.nasa.gov
conservation.org
decadeonrestoration.org
fao.org
kew.org/science
lowimpact.org
mathscareers.org
nationalgeographic.org
nhm.ac.uk
nhs.uk/live-well/eat-well
permaculturenews.org
plantheritage.org.uk
research.ed.ac.uk
rhs.org.uk/gardening-for-the-environment
sciencedaily.com
smithsonianmag.com
treezilla.org
woodlandtrust.org.uk
wwf.org.uk